Chitin- and Chitosan-Based Biocomposites for Food Packaging Applications

Chitin- and Chitosan-Based Biocomposites for Food Packaging Applications

Edited by

Jissy Jacob
Sravanthi Loganathan
Sabu Thomas

CRC Press
Taylor & Francis Group
Boca Raton London New York

CRC Press is an imprint of the
Taylor & Francis Group, an **informa** business

CRC Press
Taylor & Francis Group
6000 Broken Sound Parkway NW, Suite 300
Boca Raton, FL 33487-2742

© 2020 by Taylor & Francis Group, LLC

CRC Press is an imprint of Taylor & Francis Group, an Informa business

No claim to original U.S. Government works

Printed on acid-free paper

International Standard Book Number-13: 978-0-367-28090-1 (Hardback)

Visit the Taylor & Francis Web site at
www.taylorandfrancis.com

and the CRC Press Web site at
www.crcpress.com

Contents

List of Abbreviations

α-CTN	α-chitin
β-CTN	β-chitin
β-NAG	β-(1-4)-N-acetyl-D-glucosamine
ε-PL	ε-polylysine
A. niger	*Aspergillus niger*
ABTS•+	cationic radical 2,2′-azino-bis (3-ethylbenzothiazoline-6-sulphonic acid)
AgNPs	silver nanoparticles
AMBER	assisted model building with energy refinement
BO	born–oppenheimer
CFU	colony-forming unit
CH_4	Methane
CH_3COOH	acetic acid
CMC	carboxymethyl cellulose
CNC	cellulose nanocrystals
CNF	cellulose nanofiber
CNT	carbon nanotubes
CO_2	carbon dioxide
COMPASS	condensed-phase optimized molecular potentials for atomistic simulation studies
CS, CSN	chitosan
CSNF	chitosan-cellulose nanofiber
CSNW	chitosan-cellulose nanowhisker
$CuSO_4$	copper sulfate

DD	degree of deacetylation
DES	deep eutectic solvents
DFT	density functional theory
DMA	dynamic mechanical analyzer
DNAN	2,4-dinitroanisole
DPECS	diethoxyphosphorylpolyaminoethyl chitosan
DSC	differential scanning calorimetry
E. coli	*Escherichia coli*
EFF-MD	empirical force field molecular dynamics
EGCG	epigallocatechin gallate
$FeCl_3$	ferric chloride
f-MWCNT	functionalized multi-walled carbon nanotube
FOX7	1,1-diamino-2,2-dinitroethene
GGA	generalized gradient approximation
GlcN	N-glucosamine
GlcNAc	N-acetylglucosamine
GO	graphene oxide
GROMACS	groningen molecular simulation
HCl	hydrochloric acid
HCOOH	formic acid
HNO_3	nitric acid
HPCS	hydroxypropyl chitosan
H_2SO_4	sulfuric acid
KO	krill oil
KT	kombucha tea
LAMMPS	large-scale atomic/molecular massively parallel simulator
LDPE	low-density polyethylene
mg	Milligram
MgO	magnesium oxide
MLE	mango leaf extract
MM	molecular mechanics
MMT	montmorillonite
MOPAC	molecular orbital package
Mpa	megapascal

M_w	molecular weight
MWCNT	multiwalled carbon nanotubes
NADES	natural deep eutectic solvents
NaOH	sodium hydroxide
nm	nanometer
NTO	3-nitro-1,2,4-triazol-5-one
NQ	nitroguanidine
OL	oleic acid
OTR	oxygen transfer rate
P. aeruginosa	*Pseudomonas aeruginosa*
PAA	polyacrylic acid
PCL	polycaprolactone
PE	polyethylene
PHT	poly (3-hexylthiophene)
PM3	parameterization model version 3
PP	polypropylene
RAFT	reversible addition–fragmentation chain transfer
RDX	1,3,5-hexahydro-1,3,5-trinitro-1,3,5-triazine
S. aureus	*Staphylococcus aureus*
S. enteritidis	*Salmonella enteritidis*
SC-DFTB	self-consistent density functional tight binding
SEM	scanning electron microscopy
SiO_2	silica
SMD	steered molecular dynamics
SPIONPs	super-paramagnetic iron oxide nanoparticles
TEMPO-CNF	(2,2,6,6-tetramethylpiperidine-1-oxyl radical)-oxidized cellulose nanofibers
TEOS	tetraethoxysilane
TiO_2	titanium dioxide
TGA	thermogravimetric analyzer
TNF	turmeric nanofillers
TNT	2,4,6-trinitromethylbenzene
TPP	tripolyphosphate
USFDA	United States Food and Drug Administration
WVP	water vapor permeability
ZnO-NP	zinc oxide nanoparticles

List of Symbols

E'	storage modulus
E''	loss modulus
E_a	activation energy
T_c	crystallization temperature
T_d	decomposition temperature
T_g	glass transition temperature
T_m	melting temperature
χ_c	degree of crystallinity

Preface

Plastics are the most commonly used materials for packaging applications because of low cost, easy processability, and the availability of abundant resources for their production. The properties essential for packaging materials are determined by the physical and chemical characteristics of the product, as well as by the external conditions in which the product is stored/transported. As plastics have a wide range of properties which can be tailored according to the product requirements, they are the most attractive materials for packaging applications. Polyethylene (PE), polypropylene (PP), polyethylene terephthalate (PET), polyvinyl chloride (PVC), and polystyrene (PS) are the most common packaging plastics, accounting for more than 90% of the total volume of plastics used in packaging.

Plastics do not biodegrade—primarily because they are made of synthetic polymers and no microbe has yet evolved that can feed on them. Disposal of the millions of tons of plastic waste generated every year takes up huge areas in the form of landfills. Plastic polymers may not be toxic themselves, but the myriad of chemical monomers added to them to improve their properties can be released to their surroundings and contact materials over time or under conditions such as heat and exposure to sunlight or photo degradation. In this backdrop, the development and use of biobased and/or biodegradable polymers are gaining importance.

Chitin, is one of the most important biopolymers, synthesized by an enormous number of living organisms, and is a promising bioactive polymer for food packaging applications due to its functional properties. Chitosan is the deacetylated derivative of chitin and is nontoxic, biocompatible, and biodegradable, and is thus considered as an environmentally friendly packaging material. Moreover, chitin and chitosan are good inhibitors against the growth of a wide variety of yeasts, fungi, and bacteria, and also display gas and aroma barrier properties in dry conditions. The use of edible or biodegradable materials, plant extracts, and nanomaterials is expected to substitute synthetic additives owing to the risk they represent for the environment and human health. Therefore, this book focuses on composition, properties, characterization, and theoretical approaches to chitin and chitosan biocomposites. Also, it describes the most recent studies concerning chitin- and chitosan-based films and gives an overview about future trends regarding the industrial applications of chitin and chitosan for food packaging purposes.

The major motivation for writing this book is the environmental sustainability and use of environmentally acceptable biomaterials in the areas of scientific and industrial research. In recent times, identification and application of suitable biopolymers, instead of petroleum derived conventional polymers, are gaining more attention. This book is especially useful for researchers at various levels of academic endeavours working in the field of bionanocomposites, particularly those with an interest in packaging applications.

The editors have made a conscious effort to select authors from various parts of the world representing diverse disciplines including material science, packaging engineering, microbial sciences and food technology. We would like to thank them profusely for their high-quality submissions and for contributing to this truly multidisciplinary effort. Special thanks to our readers, and the editorial staff of CRC Press, Taylor & Francis group, for their assistance and helpful suggestions at every step.

Editors

Miss Jissy Jacob is currently working as research scholar under the guidance of Professor Sabu Thomas at Mahatma Gandhi University. She has expertise in polymer chemistry. Her research interests are in the field of biopolymer-based nanocomposites for food packaging applications. She works towards development of mesoporous materials from sustainable resources via green synthesis process and fabrication of several biopolymer composites with improved mechanical, thermal, and gas barrier properties. Her research articles are published in reputable journals.

Dr. Sravanthi Loganathan is currently working at DST INSPIRE Faculty at CSIR-Central Electrochemical Research Institute, Karaikudi, Tamil Nadu. Prior to joining CECRI, she completed her Ph.D, in Chemical Engineering from Indian Institute of Technology Guwahati, Assam, India, and worked as a Dr. Kothari Postdoctoral Fellow with Professor Sabu Thomas at School of Chemical Sciences, Mahatma Gandhi University, Kerala. Her research work is focused on biopolymer composites for food packaging and 3D printed scaffolds for tissue engineering, 3D printed monoliths based on mesoporous materials for gas separation, and sustainable composite materials for several high-performance applications. She is a recipient of several prestigious awards given by the Government of India, including Dr. Kothari

Postdoc Award, DST National Postdoc Award, DST INSPIRE Faculty Award, and Early Career Research Award. She has authored several book chapters and published several research articles in the field of porous materials for carbon capture and biopolymer-based composites for food packaging in reputable high-impact journals.

Professor Sabu Thomas is currently Vice-Chancellor of Mahatma Gandhi University, Director of School of Energy Materials, and the Founder Director and Professor of International and Interuniversity Centre for Nanoscience and Nanotechnology. He is also a full professor of Polymer Science and Engineering at School of Chemical Sciences, Mahatma Gandhi University, Kottayam, Kerala, India. Professor Thomas is an outstanding leader with sustained international acclaims for his work in nanoscience, polymer science and engineering, polymer nanocomposites, elastomers, polymer blends, interpenetrating polymer networks, polymer membranes, green composites and nanocomposites, nanomedicine, and green nanotechnology. Professor Thomas received many national and international awards and published over 800 peer-reviewed research papers, reviews, and book chapters. He has coedited 80 books and he is an inventor with six patents.

Contributors

Fauze A. Aouada
Grupo de Compósitos e
Nanocompósitos Híbridos
(GCNH), Department of
Physics and Chemistry,
Programa de Pós-
Graduação em Ciência
dos Materiais
São Paulo State University
(Unesp), School of
Engineering
Ilha Solteira, Brazil

Anuradha Biswal
Department of Chemistry
Veer Surendra Sai University
of Technology
Burla, India

Wen Shyang Chow
School of Materials and
Mineral Resources
Engineering, Engineering
Campus
Universiti Sains Malaysia
Nibong Tebal, Malaysia

Isabel Coelhoso
LAQV-REQUIMTE,
Departamento de Química,
Faculdade de Ciências e
Tecnologia
Universidade Nova de Lisboa,
Campus de Caparica
Almada, Portugal

Ana Luisa Fernando
MEtRICs, Departamento de
Ciências e Tecnologia da
Biomassa, Faculdade de
Ciências e Tecnologia, FCT
Universidade Nova de Lisboa,
Campus de Caparica
Almada, Portugal

Upendra Nath Gupta
Polymer, Petroleum and Coal
Chemistry Group,
Materials Sciences and
Technology Division
CSIR–North East Institute of
Science and Technology
Jorhat, India

Aseel T. Issa
High Point Clinical Trials
Center
High Point, NC, USA

Jayaramudu. J
Polymer, Petroleum and Coal
Chemistry Group,
Materials Sciences and
Technology Division
CSIR–North East Institute of
Science and Technology
Jorhat, India

Seema A. Kulkarni
Department of Food Process
Engineering, School of
Bioengineering
SRM Institute of Science and
Technology
Chennai, India

Mahesh Kumar. M
Department of Food Process
Engineering, School of
Bioengineering
SRM Institute of Science and
Technology
Chennai, India

Marcia R. de Moura
Grupo de Compósitos
e Nanocompósitos
Híbridos (GCNH),
Department of Physics and

Chemistry, Programa de
Pós-Graduação em Ciência
dos Materiais
São Paulo State University
(Unesp), School of
Engineering
Ilha Solteira, Brazil

João Ricardo Afonso Pires
MEtRICs, Departamento de
Ciências e Tecnologia da
Biomassa, Faculdade de
Ciências e Tecnologia, FCT
Universidade Nova de Lisboa,
Campus de Caparica
Almada, Portugal

Maria Rapa
University Politehnica of
Bucharest
Bucharest, Romania

Carolina Rodrigues
MEtRICs, Departamento de
Ciências e Tecnologia da
Biomassa, Faculdade de
Ciências e Tecnologia, FCT
Universidade Nova de Lisboa,
Campus de Caparica
Almada, Portugal

Periyar Selvam. S
Department of Food Process
Engineering, School of
Bioengineering

SRM Institute of Science and
    Technology
Chennai, India

Emmanuel Rotimi Sadiku
Department of Chemical,
    Metallurgical and Materials
    Engineering
Tshwane University of
    Technology
Pretoria, Republic of
    South Africa

K. P. Sajesha
School of Chemical Sciences
Mahatma Gandhi University
Kottayam, India

**Victor Gomes Lauriano
    Souza**
MEtRICs, Departamento de
    Ciências e Tecnologia da
    Biomassa, Faculdade de
    Ciências e Tecnologia,
    FCT
Universidade Nova de Lisboa,
    Campus de Caparica
Almada, Portugal

Sarat K. Swain
Department of Chemistry
Veer Surendra Sai University
    of Technology
Burla, India

Reza Tahergorabi
Food and Nutritional Sciences
    Program
North Carolina Agricultural
    and Technical State
    University
Greensboro, NC, USA

V. Dharini
Department of Food Process
    Engineering, School of
    Bioengineering
SRM Institute of Science and
    Technology
Chennai, India

Cornelia Vasile
Physical Chemistry of
    Polymers Department
"Petru Poni" Institute of
    Macromolecular Chemistry
Iaşi, Romania

CHAPTER 1

An Overview of Biopolymers from Natural Resources

Marcia R. de Moura and Fauze A. Aouada

*Grupo de Compósitos e Nanocompósitos Híbridos (GCNH),
Department of Physics and Chemistry, Programa de
Pós-Graduação em Ciência dos Materiais
São Paulo State University (Unesp), School of Engineering
Ilha Solteira, Brazil*

1.1 INTRODUCTION

Biopolymers can be defined as any polymers (such as proteins and polysaccharides) originating from living organisms. They are mainly produced from carbon sources, usually carbohydrates obtained from large-scale commercial plants such as sugarcane, corn, potatoes, wheat, and beets—or other vegetables, for instance, the oil extracted from soybeans, or sunflower, palm, or other oleaginous plants (Steinbüchel 2003). Some biopolymers' origins are shown in Figure 1.1.

Biopolymers are promising candidates for use as matrices for packaging materials due to their exceptional capability of film-forming, biological properties, and excellent physicochemical

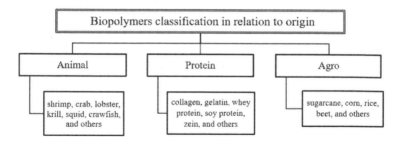

FIGURE 1.1 Biopolymer Classification: Animal, Agro, and Protein Origins.

properties (Wu et al. 2019). The application of biopolymers obtained from natural resources in food packaging is more promising. This is owing to the fact that these biopolymers are affordable, available in abundance, and renewable in nature (Shahabi-Ghahfarrokhi, Goudarzi, and Babaei-Ghazvini 2019).

1.2 PETROLEUM-BASED POLYMERS—CHALLENGES

In current society, dependence on synthetic polymers has increased over the years. Polyethylene (PE) and polypropylene (PP) are polymers obtained from polymerization reactions by using monomers obtained from petroleum derivatives (Pathak and Navneet 2017). In general, people use this polymeric class because of their desirable properties such as excellent mechanical properties, resistivity, and low chemical reactivity (Song, Murphy, Narayan and Davies 2009; Dey, Mondal, Das and Dutta 2012). However, some problems arise from the use of synthetic polymers. Their low response against different types of chemical/ biological degradation creates a problem from the moment they are discarded. It is estimated that a simple plastic bag that one takes home from the supermarket requires about 450 years to degrade.

In recent years, the use of synthetic polymers has presented problems in relation to the crisis in the acquisition of petroleum resources, mainly in relation to cost, as well as an environmental concern.

Thus, biodegradable polymers have been a viable alternative to minimize this environmental impact. These materials began to be studied in the mid-1960s. Thus, bio-based matrices have been widely studied in order to replace synthetic polymers.

1.3 BIOPOLYMERS

One main advantage of substitution of biopolymers for petroleum-based polymers is that they are totally capable of biodegradation and break down into the simplest molecules, such as CO_2, H_2O, or CH_4. This can occur by an enzymatic action initiated by microorganisms present in the environment. Polymeric matrices obtained from renewable resources may present biodegradability or may be compostable under controlled conditions. In industry, the application of biopolymers can be found in the most diverse areas. For instance, biopolymers are used to produce smart textiles using eco-friendly manufacturing processes.

1.4 CLASSIFICATION

Biopolymer fibers may classified as natural, regenerated, or synthetic and their use can contribute to reducing environmental pollution (Younes 2017). Biopolymers such as polycaprolactone (PCL) can be processed as spun fibers or manufactured to be applied in soft tissue engineering, or as electro-spun fibers and seat belts (Asadian et al. 2019). Natural biopolymer-based films or their nanocomposites have been studied extensively for packaging applications in the food industry (de Moura et al. 2009; Otoni et al. 2017; Nunes, Melo, Aouada, and de Moura 2018). Biopolymers based on hydrogels containing drugs have shown positive use in drug delivery (Fernandes, de Moura, Glenn, and Aouada 2018). Various other preparation methods have been used in the synthesis of biopolymeric hydrogels for possible applications in agriculture (Ferreira Junior, de Moura, and Aouada 2017; Tanaka, Ferreira Junior, de Moura, and Aouada 2018).

1.4.1 Agro Origin

Several authors described the extraction of polysaccharides from agroindustrial wastes. For instance, the extraction of both chitin and chitosan polysaccharides by *Cunninghamella elegans* and *Rhizopus arrhizus* strains from corn steep liquor– and molasses-based agroindustrial wastes was reported by Berger et al. (2014). Leite et al. (2015) described yet another effective route for extraction of chitosan from agroindustrial substrates such as corn steep liquor and sugar cane bagasse. The steps involved in the extraction process include deproteinization of agroindustrial wastes using NaOH, which is then followed by autoclaving and centrifugation. It is reported that the degree of deacetylation of chitosan is achieved to be ~80%. Additionally, the authors also documented that the chitosan acts as a fungicide against pathogenic yeasts. Hence, it is possible to expect that chitosan can be tested as a component of food packaging.

The main advantage of the use of chitosan from agroindustrial wastes in packaging over conventional materials, such as synthetic polymers is to reduce environmental pollution. It has low cost when compared to other synthetic materials. In addition, when used in the nanoparticulate state, it may act as an efficient additive for improving the physicochemical and barrier properties related to food packaging (Ramesh and Radhakrishnan 2019).

1.4.2 Animal Origin

Chitin is the second most plentiful polysaccharide on the Earth. Due to high crystallinity of strong hydrogen bonds and cohesive forces, chitin is practically insoluble in all regular solvents, including water and organic substances, affecting expansion of the processes for the preparation of chitin-based derivatives (Roy, Salaün, Giraud, and Ferri 2017). For instance, Younes and Rinaudo (2015) described various methods for recovery chitin from different marine organisms, including shrimp, crabs, lobsters, krill, squid, and crawfish, using NaOH concentration, temperature, and time as deproteinization experimental parameters, and HCl concentration, temperature, and time as factors in the demineralization process. According to the authors, chitosan is frequently obtained from chitin deacetylation. For the degree of acetylation that is inferior to 0.5, chitosan becomes soluble in several solvents, such as the aqueous acidic medium. Because of these solubility properties, several works have demonstrated the application of chitosan biopolymers in food packaging. Among different sources, those of agro, animal, and protein origin are the three main sources of obtaining biopolymers.

Chitosan biopolymers of animal origin are mainly found in animals lacking a backbone, such as arthropods, various insects and marine diatoms, and algae (Synowiecki and Al-Khateeb 2003). For instance, Islam, Khan, and Nowsad Alam (2016) extracted chitin from shrimp shell wastes and produced chitosan on a commercial scale from alkaline hydrolysis using NaOH. Recently, Hugo and Hugo (2015) presented a complete revision about the application of different animal-derived chitosans as natural preservatives in fresh sausage products. This study showed that chitosan acts against different microorganisms, increases the shelf life and reduces the lipid oxidation of boerewors and skinless pork sausages, and improves the quality of these products by acting as a food packing. According to Bilbao-Sainz et al. (2018), due to their interesting properties such as biodegradability, biocompatibility, and non-toxicity, chitosan is a polysaccharide with the potential to be applied in food coating. In this work, the authors produced coating using fungal chitosan and alginate by self-assembly. They found that the polysaccharide coating decreased the yeasts and fungal evolution, and improved the fruit bar shelf life, preserving the quality of these food products.

1.4.3 Protein Origin

The literature reports several composites, biocomposites, and nanocomposites formed from protein-chitosan films for food use. Collagen/chitosan composites plasticized by using glycerol have great potential for applications in this area because it presented satisfactory mechanical, transparency, solubility, and thermal properties (Ahmad et al. 2016).

Different strategies have been employed to improve the chitosan-based product characteristics. Qu et al. (2019) observed that the TiO_2 nanoparticles increased the tensile force of the zein/chitosan nanocomposites, reaching around 28 MPa at 0.15 wt % TiO_2 nanoparticles. It was also observed that these nanocomposites had an antibacterial effect on gram-positive bacteria (*S. aureus*) and gram-negative bacteria (*E.coli* and *S. enteritidis*), and this effect was made more significant by UV light conditions. Samsi et al. (2019) first reported the use of gelatin-chitosan films in the preservation of cherry tomatoes and grapes. The addition of chitosan increased mechanically, specifically with respect to transparency, vapor, and water permeability. In addition, the antimicrobial properties of these films were similar to those of the commercial cling films.

1.5 BIONANOCOMPOSITES

An alternative for the improvement of the properties of biodegradable biopolymers is the development of nanocomposites with the use of nanoparticles of reinforcements (or nanofillers) that are also from the renewable and biodegradable source.

A bionanocomposite is a matrix formed from two or more constituents (as mentioned, one being from living organisms) that form new materials with better performance during individual phases. These phases are the continuous phase (or matrix) and reinforcing phase. In general, the reinforcing phase is formed by fibers, whiskers, and various particle types (Rudin and Choi 2013). The matrix binds the reinforcing

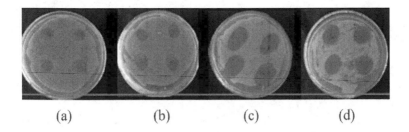

(a)   (b)   (c)   (d)

FIGURE 1.2 Inhibitory Effects of Silver Nanoparticles and Alginate Films versus Gram-Positive Bacteria *S. aureus* (A) and (B) and Gram-Negative *E. coli* (C) and (D).

phases together and offers the mechanical sustenance. One application is the introduction of silver nanoparticles into natural polymers for use as curatives. It has already been proved that nanoparticles potentiate the efficiency of silver against bacteria (Figure 1.2).

ACKNOWLEDGMENTS

The authors thank Universidade Estadual Paulista, and Fundação de Amparo à Pesquisa do Estado de São Paulo (FAPESP) and Conselho Nacional de Desenvolvimento Científico e Tecnológico (CNPq) for financial support. This study was financed in part by the Coordenação de Aperfeiçoamento de Pessoal de Nível Superior—Brasil (CAPES)—Finance Code 001".

REFERENCES

Ahmad, M., Nirmal, N. P., Danish, M., Chuprom, J., Jafarzedeh, S. 2016. Characterisation of composite films fabricated from collagen/chitosan and collagen/soy protein isolate for food packaging applications. *RSC Adv* 6:82191–82204.

Asadian, M., Onyshchenko, J., Thiry, D. et al. 2019. Thiolation of polycaprolactone (PCL) nanofibers by inductively coupled plasma (ICP) polymerization: Physical, chemical and biological properties. *Appl Surf Sci* 479:942–952.

Berger, L. R. R., Stamford, T. C. M., Stamford-Arnaud, T. M. et al. 2014. Green conversion of agroindustrial wastes into chitin and chitosan by *Rhizopus arrhizus* and *Cunninghamella elegans* strains. *Int J Mol Sci* 15:9082–9102.

Bilbao-Sainz, C., Chiou, B., Punotai, K. et al. 2018. Layer-by-layer alginate and fungal chitosan-based edible coatings applied to fruit bars. *J Food Sci* 83:1880–1887.

Dey, U., Mondal, N. K., Das, K., Dutta, S. 2012. An approach to polymer degradation through microbes. *J Pharm* 2:385–388.

Fernandes, R. S., de Moura, M. R., Glenn, G. M., Aouada, F. A. 2018. Thermal, microstructural, and spectroscopic analysis of Ca^{2+} alginate/clay nanocomposite hydrogel beads. *J Mol Liq* 265:327–336.

Ferreira Junior, C. R., de Moura, M. R., Aouada, F. A. 2017. Synthesis and characterization of intercalated nanocomposites based on poly(methacrylic acid) hydrogel and nanoclay cloisite-Na for possible application in agriculture. *J Nanosci Nanotechnol* 17:5878–5883.

Hugo, C. J., Hugo, A. 2015. Current trends in natural preservatives for fresh sausage products. *Trends Food Sci Technol* 45:12–23.

Islam, S. Z., Khan, M., Nowsad Alam, K. M. 2016. Production of chitin and chitosan from shrimp shell wastes. *J Bangladesh Agril Univ* 14:253–259.

Leite, M. V., Stamford, T. C. M., Stamford-Arnaud, T. M. et al. 2015. Conversion of agro-industrial wastes to chitosan production by *Syncephalastrum racemosum* UCP 1302. *Int J Appl Res Nat Prod* 8:5–11.

de Moura, M. R., Aouada, F. A., Avena-Bustillos, R. J. et al. 2009. Improved barrier and mechanical properties of novel hydroxypropyl methylcellulose edible films with chitosan/tripolyphosphate nanoparticles. *J Food Eng* 92:448–453.

Nunes, J. C., Melo, P. T. S., Aouada, F. A., de Moura, M. R. 2018. Influência da nanoemulsão de óleo essencial de limão em filmes à base de gelatina. *Quim Nova* 41:1006–1010.

Otoni, C. G., Avena-Bustillos, R. J., Azeredo, H. M. C. et al. 2017. Recent advances on edible films based on fruits and vegetables—A review. *Compr Rev Food Sci Food Saf* 16:1151–1169.

Pathak, V. M., Navneet. 2017. Review on the current status of polymer degradation: A microbial approach. *Bioresour Bioprocess* 4:15.

Qu, L., Chen, G., Dong, S. et al. 2019. Improved mechanical and antimicrobial properties of zein/chitosan films by adding highly dispersed nano-TiO_2. *Ind Crops Prod* 130:450–458.

Ramesh, S., Radhakrishnan, P. 2019. Cellulose nanoparticles from agro-industrial waste for the development of active packaging. *Appl Surf Sci* 484:1274–1281.

Roy, J. C., Salaün, F., Giraud, S., Ferri, A. 2017. *Solubility of Chitin: Solvents, Solution Behaviors and Their Related Mechanisms.* London: InTechOpen.

Rudin, A., Choi, P. 2013. *The Elements of Polymer Science and Engineering.* (Third Edition). London: Elsevier.

Samsi, M. S., Kamari, A., Fatimah, I., Sumardi, Yusoff, S. N. M. 2019. Synthesis, characterisation and application of gelatin-chitosan blend films for fruit preservation. *Fresen Environ Bull* 28:30–43.

Shahabi-Ghahfarrokhi, I., Goudarzi, V., Babaei-Ghazvini, A. 2019. Production of starch based biopolymer by green photochemical reaction at different UV region as a food packaging material: Physicochemical characterization. *Int J Biol Macromol* 122:201–209.

Song, J. H., Murphy, R. J., Narayan, R., Davies, G. B. H. 2009. Biodegradable and compostable alternatives to conventional plastics. *Philos Trans R Soc Biol* 364:2127–2139.

Steinbüchel, A. 2003. *Biopolymers Vol. 10: General Aspects and Special Applications* (Vol. 10). Hoboken: Wiley-VCH.

Synowiecki, J., Al-Khateeb, N. A. 2003. Production, properties, and some new applications of chitin and its derivatives. *Crit Rev Food Sci Nutr* 43:145–171.

Tanaka, F. N., Ferreira Junior, C. R., de Moura, M. R., Aouada, F. A. 2018. Water absorption and physicochemical characterization of novel zeolite-PMAA-co-PAAm nanocomposites. *J Nanosci Nanotechnol* 18:7286–7295.

Wu, C., Zhu, Y., Wu, T. et al. 2019. Enhanced functional properties of biopolymer film incorporated with curcurmin-loaded mesoporous silica nanoparticles for food packaging. *Food Chem* 288:139–145.

Younes, B. 2017. Classification, characterization, and the production processes of biopolymers used in the textiles industry. *J Text I* 108:674–682.

Younes, I., Rinaudo, M. 2015. Chitin and chitosan preparation from marine sources. Structure, properties and application. *Mar Drugs* 13:1133–1174.

CHAPTER 2

Fabrication of Bionanocomposites from Chitin

Seema A. Kulkarni, V. Dharini, S. Periyar Selvam, and M. Mahesh Kumar

Department of Food Process Engineering, School of Bioengineering
SRM Institute of Science and Technology
Chennai, India

Emmanuel Rotimi Sadiku

Department of Chemical, Metallurgical and Materials Engineering
Tshwane University of Technology
Pretoria, Republic of South Africa

J. Jayaramudu and Upendra Nath Gupta

Polymer, Petroleum and Coal Chemistry Group, Materials Sciences
* and Technology Division*
CSIR–North East Institute of Science and Technology
Jorhat, India

2.1 INTRODUCTION

Next to cellulose, chitin (CTN) is the richest natural mucopolysaccharide in the world (Kumar 2000). It occurs natively as crystalline microfibrils in an ordered form that imparts strength to the exoskeletons of arthropods, as well as the outer shells of insects and crustaceans, and also cell walls of fungi. Currently, crab and shrimp shells are the only two sources that are utilized extensively for extraction, although extraction is possible using many other sources. In industrial applications, CTN is extracted by acid treatment for the purpose of decolorization, demineralization, and deproteinization. Presence of hydrophobic acetyl groups renders chitin insoluble in water and some organic solvents, which poses a major problem for its characterization and use. However, this can be overcome by transforming it into chitosan (CSN) via partial deacetylation when it is subjected to alkaline conditions. The chemical structures for chitin and chitosan are shown in Figures 2.1 and 2.2, respectively. CTN and related polymer studies have gained tremendous momentum over the past few decades owing to their versatility with high potential towards applications in various fields of technology namely, food

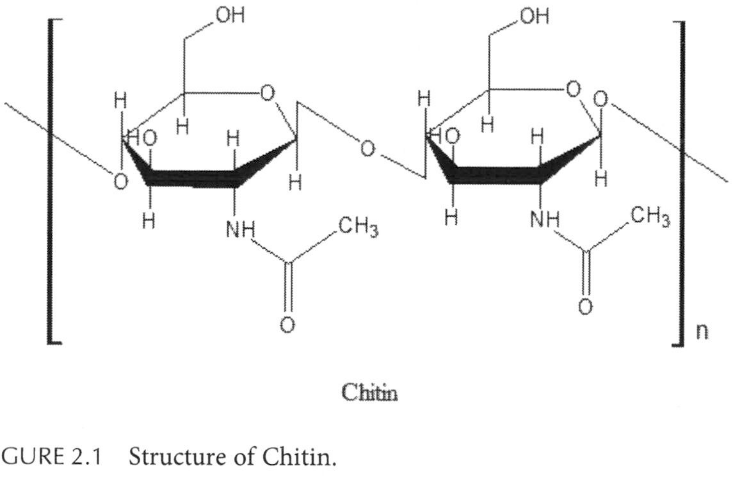

Chitin

FIGURE 2.1   Structure of Chitin.

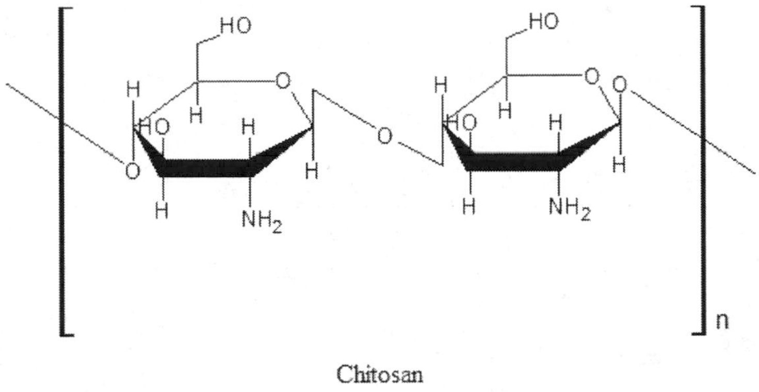

Chitosan

FIGURE 2.2   Structure of Chitosan.

packaging, agriculture, cosmetics, biopharmaceuticals, paper, textiles, biomedicine, sewage treatment, as well as heavy metal chelation and photographic products (Aranaz et al. 2018).

2.2 SOURCES OF CHITIN

CTN is extensively dispensed in nature, and it is considered to be one of the most important renewable resources. It is found in cuticles or exoskeletons of invertebrates and in the cell walls of algae and fungi species (Muzzarelli and Pariser 1978). CTN obtained from squid pens and Antarctic krill species waste is important in its practical utilization. Dry weight of chitin obtained from crustaceans are highest when compared with the other sources. Arthropods from lakes, ponds, and rivers are the major species contributing to the production of chitin. A clear-cut dominance in CTN production is observed from zooplankton, most particularly holoplanktonic crustaceans. CTN produced from crustaceans such as crayfish, spiny lobsters, and squillae is currently of industrial interest. CTN content from different sources varies in the range from 7% for barnacles to 40% for squid pens. Species that are taxonomically related have very similar CTN content and this can be observed in red marbled crabs as well as in spider crabs. Squids and crayfish have almost indistinguishable amounts of CTN, ranging between 35% and 40%, and this is considered to be one of the interesting areas of research (Rhazi et al. 2000).

2.3 STRUCTURE AND COMPOSITION OF CHITIN

CTN is a large, structural homopolymer of β-(1-4)-N-acetyl-D-glucosamine (β-NAG) linkages. Based on source of origin, it occurs as two allomorphs, namely, α- and β-forms (Rinaudo 2006). The chains of chitin are interlinked via hydrogen bonding, which impacts its solubility, reactivity, and swelling (Kayra and Aytekin 2019). The α chitin isomorph is abundant in nature, as compared to β counterpart (Rajkumari and Busi 2018). The crystallographic measures of the two isomorphs clearly display the fact that per unit cell in α-chitin (α-CTN), there are two antiparallel molecules, but only one in β-chitin (β-CTN), which are arranged in parallel. In both cases, chains in the sheets are held via intrasheet hydrogen bonds, which are assumed to account for the resistance with respect to swelling of α-CTN in water.

2.4 PROPERTIES OF CHITIN

CTN is a naturally occurring structural polysaccharide, which is white in color, hard, and inelastic. Unlike other natural polysaccharides, which are either neutral or acidic, CTN is highly basic. Besides typically being hydrophobic and insoluble in water, it has a chelating property, the ability to form polyoxysalt and films. It is soluble in certain organic solvents such as hexafluoroacetone, hexafluoro-isopropanol, and chloroalcohols along with aqueous solutions of mineral acid and dimethylacetamide containing 5% lithium chloride (Cho et al. 2000). CSN is soluble in weak acids, e.g., formic acid. Neither CTN nor CSN have any well-characterized melting points, but they char on excessive application of heat. In CTN, the water molecules are tightly bound and hence, water loss generally occurs in the temperature regime of 53° to 100°C. The energy required to vaporize the water molecules present in chitin is almost ten times higher than that of the energy necessary to carry out vaporization of an equivalent quantity of free water (Neto et al. 2005). Both CTN and CSN have numerous advantageous biological properties. They are hemostatic, fungistatic, and spermicidal; they also bind well to microbial and mammalian cells, regenerate gum tissue, and accelerate the synthesis of osteoblasts that is responsible for bone formation. They exhibit antitumor, antioxidant, and anticholesteremic properties. They can also act as central nervous system depressants and immunologic adjuvants. They have applications in multiple streams because of their fiber- and film-forming properties.

2.5 ISOLATION AND CHARACTERIZATION OF CHITIN

Commercial treatment in the production of CTN usually involves harsh acids and alkalis, which discharge many harmful chemical wastes into the environment. Biological process is used as an alternative approach which provides cheaper production cost and is eco-friendly. The bacterial strains isolated from soil and sediment samples are capable of degrading chitin and exhibit antifungal activity for plant diseases (Hoster, Schmitz, and Daniel 2005). Nanowhiskers from CTN can be produced with the length and width of ~100 nm and ~10 nm, respectively, by acid hydrolysis of fermented prawn waste (Asri et al. 2017). Apart from traditional sources, CTN can be isolated from such nonconventional sources as insects and other fungal mycelia. CTN isolated from bumblebee corpses can be used for the homogenous dispersion of pectin and alginate with excellent physio-mechanical properties. Mild demineralization treatment is necessary for the recovery of chitin, as insect corpses do not contain inorganic material (Majtán et al. 2007). To attain uniformity of the product, the relationship between the processing conditions and characteristics of the products should be monitored continuously. This perspective can produce products with assorted grades based on the intended usage.

2.6 MODIFICATIONS

CTN has been underutilized due to certain properties that hinder their practical applications. Therefore, many types of modifications have been tried and tested in order to explore new properties. Chemical modification is preferred since it does not interfere with physicochemical and biochemical properties and the fundamental skeletons of CTN and CSN. Numerous methods have been documented for some time, and a list of a few of these methods follows:

1. Sulphated CTN and CSN: Since sulphate acts as the closest structural analogue of heparin, a natural blood anticoagulant, the amino and hydroxyl groups in CTN and CSN are specifically sulphated for particular applications in the field of medicine, namely, anticoagulant, antisclerotic, and antiviral activities (Jayakumar et al. 2007).

2. Chemically modified chitin with polypyrrole: Application of this modification for eliminating heavy metal contamination, such as lead and cadmium ions from aqueous solutions (Karthik and Meenakshi 2015).

3. Phosphorylated CTN and CSN: Phosphorylation can be achieved by several grafting and reactive methods. They exhibit antibacterial, bioabsorbable, osteoinductive, and metal chelating properties (Jayakumar et al. 2008).

4. Hydrophobic modification of CTN whiskers: It is basically the reaction of chitin whiskers with bromohexadecane in order to hydrophobically modify the whiskers and evaluate their performance in oil thickening and structuring (Huang et al. 2015).

5. Surface modification of CTN and CSN via oxidative polymerization: This method involves the grafting of poly (3-hexylthiophene) [PHT] to the surfaces of CTN and CSN directly via oxidative polymerization with $FeC\ell_3$. This

process upgrades the electrical property and also converts hydrophilic surface behavior of CTN and CSN to hydrophobic, which is evident from the water contact angles of ~97.7° and ~107.0°, respectively, in the presence of PHT (Hai and Sugimoto 2018).

2.7 CHITIN BIOCOMPOSITES

Natural polymers are vital in the development of biocomposite materials for industrial applications such as renewability, biodegradability, and low carbon release, low cost, non-petroleum–based sources (Moon et al. 2011). Composite materials are an assembly of at least two immiscible components (Continuous phase matrix and the reinforcement). The main function of reinforcement is to improve the mechanical, thermal, and biological properties of the matrix. The distinct properties of chitin are utilized in packaging, coating, water treatment, biomedicine, and in other fields and areas. In order to increase the mechanical and water resistant characteristics of biocomposite materials, certain nano-fillers are incorporated in the matrix. The use of graphene oxide (GO) as nano- reinforcement in polymer matrices often results in low uptake of moisture and improved water resistance of the biocomposite films (Ma et al. 2012). The CTN-based biocomposite, comprised of 60% biopolymer, 40% polymer (Polyurethane), and 2% catalyst (Hydroxyl and tertiary amine), is reported to adsorb 0.29 mg of fluoride per gram and is hence concluded to be useful for removal of wide range of contaminants from water (Davila-Rodriguez et al. 2009). Recycling of thermoplastic biocomposites is done by the molding technique. Melting temperature is one of the restrictive conditions in recycling of chitin-based composites and it should not increase beyond 300°C, at which point it begins to degrade. This technique does not damage the reinforcement in the biocomposite material. CTN has attracted significant and notable attention in the last five years as a biodegradable material with a satisfactory way of waste management of the crustacean shells.

2.8 APPLICATIONS IN PACKAGING INDUSTRIES

CTN biopolymers are both biodegradable and renewable. With the purpose of achieving the preferred functionalities in order to improve the product safety, an active packaging system is employed in which active components are loaded in the chitin matrix (Demitri et al. 2016). These active agents can act as absorbers, releasers, or reacting agents. It is stated that the global market for biobased plastics was less than 1% in 2009 and is anticipated to increase to ~20% by the end of 2020. Solubility of chitin in acidic media is a valuable characteristic, since it can be modified as fibers, films, sponges, beads, gels, or solutions and blended with polymers of both natural and synthetic origin. Nanocapsules can be generated by using electrospraying techniques, which produce ultrathin solids with health-promoting active packaging strategies. The CTN/titanium oxide (TiO_2) composite-based packaging system showed enhanced improvement in wettability, as well as in mechanical characteristics, and helped in perseveration of grapes from microbial attack (Zhang et al. 2017). There are numerous challenges as well as opportunities involved in commercializing the biodegradable packaging. An extra major requirement for such biocomposite films used in packaging applications is that they comply with the government legislation. To be more specific, such packaging material should exhibit minimal or no migration effect and not interact with the product. Biopackaging is expected to offer a clean and pollution-free environment in the future.

ACKNOWLEDGMENTS

The author, Emmanuel Rotimi Sadiku, wishes to acknowledge the financial support by The National Institute for the Humanities and Social Sciences (NIHSS) BRICS Research and Teaching Mobility Grant 2019, African Pathways Programme (APP) (BMG19/1013).

REFERENCES

Aranaz, Inmaculada, Niuris Acosta, Concepción Civera, Begoña Elorza, Javier Mingo, Carolina Castro, María Gandía, and Angeles Heras Caballero. 2018. "Cosmetics and cosmeceutical applications of chitin, chitosan, and their derivatives." *Polymers* no. 10 (2):213.

Asri, Syazeven Effatin Azma Mohd, Zainoha Zakaria, Reza Arjmandi, Azman Hassan, and M.K. Mohamad Haafiz. 2017. "Isolation and characterization of chitin nanowhiskers from fermented tiger prawn waste." *Chemical Engineering Transactions* no. 56:139–144.

Cho, Yong-Woo, Jinho Jang, Chong Rae Park, and Sohk-Won Ko. 2000. "Preparation and solubility in acid and water of partially deacetylated chitins." *Biomacromolecules* no. 1 (4):609–614.

Davila-Rodriguez, Jose L., Vladimir A. Escobar-Barrios, Keiko Shirai, and Jose R. Rangel-Mendez. 2009. "Synthesis of a chitin-based biocomposite for water treatment: Optimization for fluoride removal." *Journal of Fluorine Chemistry* no. 130 (8):718–726.

Demitri, Christian, Vincenzo Maria De Benedictis, Marta Madaghiele, Carola Esposito Corcione, and Alfonso Maffezzoli. 2016. "Nanostructured active chitosan-based films for food packaging applications: Effect of graphene stacks on mechanical properties." *Measurement* no. 90:418–423.

Hai, Thien An Phung, and Ryuichi Sugimoto. 2018. "Surface modification of chitin and chitosan with poly (3-hexylthiophene) via oxidative polymerization." *Applied Surface Science* no. 434:188–197.

Hoster, Frank, Jessica E. Schmitz, and Rolf Daniel. 2005. "Enrichment of chitinolytic microorganisms: Isolation and characterization of a chitinase exhibiting antifungal activity against phytopathogenic fungi from a novel Streptomyces strain." *Applied Microbiology and Biotechnology* no. 66 (4):434–442.

Huang, Yao, Meng He, Ang Lu, Weizheng Zhou, Simeon D. Stoyanov, Eddie G. Pelan, and Lina Zhang. 2015. "Hydrophobic modification of chitin whisker and its potential application in structuring oil." *Langmuir* no. 31 (5):1641–1648.

Jayakumar, Rangasamy, Nitar Nwe, Seiichi Tokura, and Hifumi Tamura. 2007. "Sulfated chitin and chitosan as novel biomaterials." *International Journal of Biological Macromolecules* no. 40 (3):175–181.

Jayakumar, Rangasamy, Nagarajan Selvamurugan, Shanti kumar V. Nair, Seiichi Tokura, and Hifumi Tamura. 2008. "Preparative methods of phosphorylated chitin and chitosan – An overview." *International Journal of Biological Macromolecules* no. 43 (3):221–225.

Karthik, Rathinam, and Sankaran Meenakshi. 2015. "Chemical modification of chitin with polypyrrole for the uptake of Pb (II) and Cd (II) ions." *International Journal of Biological Macromolecules* no. 78:157–164.

Kayra, Neslihan, and Ali Özhan Aytekin. 2019. "Synthesis of cellulose-based hydrogels: Preparation, formation, mixture, and modification." In *Cellulose-Based Superabsorbent Hydrogels. Polymers and Polymeric Composites: A Reference Series*, ed. Mondal Md. Ibrahim H., 407–434. Springer, Cham.

Kumar, Majeti N.V. Ravi. 2000. "A review of chitin and chitosan applications." *Reactive and Functional Polymers* no. 46 (1):1–27.

Ma, Jun, Changhua Liu, Rui Li, and Jia Wang. 2012. "Properties and structural characterization of chitosan/graphene oxide biocomposites." *Bio-Medical Materials and Engineering* no. 22 (1–3):129–135.

Majtán, Juraj, Katarína Bíliková, Oskar Markovič, Ján Gróf, Grigorij Kogan, and Jozef Šimúth. 2007. "Isolation and characterization of chitin from bumblebee (*Bombus terrestris*)." *International Journal of Biological Macromolecules* no. 40 (3):237–241.

Moon, Robert J., Ashlie Martini, John Nairn, John Simonsen, and Jeff Youngblood. 2011. "Cellulose nanomaterials review: Structure, properties and nanocomposites." *Chemical Society Reviews* no. 40 (7):3941–3994.

Muzzarelli, Riccardo A.A., and Ernst R. Pariser. 1978. "Proceedings of the First International Conference on Chitin/Chitosan: [held in Boston, Mass. on April 11 through 13, 1977]." *United States. National Technical Information Service. PB (USA)*.

Neto, Cypriano Galvão da Trindade, José Alberto Giacometti, Aldo Eloizo Job, Fábio Cop Ferreira, José Luís Cardozo Fonseca, and Márcia Rodrigues Pereira. 2005. "Thermal analysis of chitosan based networks." *Carbohydrate Polymers* no. 62 (2):97–103.

Rajkumari, Jobina, and Siddhardha Busi. 2018. "Advances in biomedical application of chitosan and its functionalized nano-derivatives." In *Fungal Nanobionics: Principles and Applications*, eds. Prasad Ram, Kumar Vivek, Kumar Manoj, and Wang Shanquan, 145–163. Springer, Singapore.

Rhazi, Mohammed, Jacques Desbrieres, Abdelouahad Tolaimate, Abdelhakim Alagui, and P. Vottero. 2000. "Investigation of different natural sources of chitin: Influence of the source and deacetylation process on the physicochemical characteristics of chitosan." *Polymer International* no. 49 (4):337–344.

Rinaudo, Marguerite. 2006. "Chitin and chitosan: Properties and applications." *Progress in Polymer Science* no. 31 (7):603–632.

Zhang, Xiaodong, Gang Xiao, Yaoqiang Wang, Yan Zhao, Haijia Su, and Tianwei Tan. 2017. "Preparation of chitosan-TiO$_2$ composite film with efficient antimicrobial activities under visible light for food packaging applications." *Carbohydrate Polymers* no.169:101–107.

CHAPTER 3

Fabrication of Bionanocomposites from Chitosan

Preparation to Applications

Anuradha Biswal and Sarat K. Swain

Department of Chemistry
Veer Surendra Sai University of Technology
Burla, India

3.1 INTRODUCTION

Chitin is an ecologically safe, renewable, biodegradable, biocompatible, and biofunctional substance. Therefore, it is deemed suitable for packaging applications. However, chitin is sparingly soluble and infusible during chemical treatments. The issue of its insolubility poses a major drawback in the processing, development, and characterization of this biopolymer. To address this, chitin is modified into chitosan, which is an N-deacetylated form of chitin. Due to its solubility in water, chitosan is conveniently used for research purposes (Revathi, Saravanan, and Shanmugam 2012). Chitosan, a linear polysaccharide and deacetylated derivative of chitin, consists of *N*-acetyl-*D*-glucosamine and *D*-glucosamine units connected by β (1,4) glycosidic linkages. The schematic representation involved in extraction of chitosan is presented in Figure 3.1. It is a nontoxic, polycationic, biodegradable, biocompatible, and commercially available polycationic biopolymer containing positive charges due to the presence of amino groups (Anaya et al. 2013). Chitosan is successfully derived by alkaline and enzymatic methods from chitin, primarily collected from crustacean shells (Venkatesan and Kim 2010).

Both chitin and chitosan are outstanding candidates for wide-scale applications in food, agriculture, cosmetics, and pharmaceutical industries (Demitri et al. 2016). They possess inherent antimicrobial properties which have opened them for various applications in packaging fields (Tan et al. 2015). In addition to the inherent antimicrobial attributes, the application of chitosan to provide a protective covering for

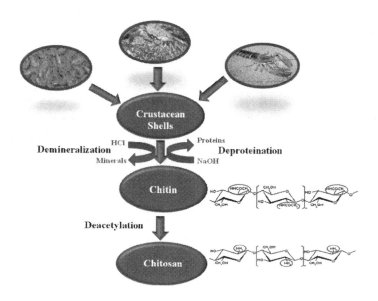

FIGURE 3.1 The Schematic Representation Involved in Extraction of Chitosan.

food preservation has led to improvement in food sensory characteristics (Souza et al. 2011).

The physicochemical characteristics of chitosan can be enhanced drastically by chemical modification or by compositing it through incorporation of some reinforcing agents. These modifications result in transformation of mechanical strength, antimicrobial nature, biocompatibility and biodegradability, tensile strength, and barrier properties. Furthermore, certain active agents can be incorporated into chitosan matrix to form composite materials that allow slow and sustained release of active agents. This provides effective protection against microbial growth that causes food degradation (Ferraro et al. 2010). This chapter discusses the various modes of isolation of chitosan from natural sources and the formation of various chitosan biocomposites. A detailed analysis is provided of the characterization, properties, and applications of chitosan biocomposites in packaging.

3.2 STRUCTURE AND COMPOSITION OF CHITOSAN

Chitosan, a natural, linear, amino polysaccharide contains copolymers of N-acetylated and deacetylated units of glucosamine joined by β (1, 4) glycosidic linkages. It is procured from partial deacetylation of chitin (Anaya et al. 2013). The ratio of glucosamine/N-acetyl-D-glucosamine gives the degree of deacetylation. If the percentage of N-acetyl glucosamine is higher, the resulting polymer is called chitin. On the other hand, if the percentage of the deacetylated unit is higher, the polymer is termed chitosan (Ramírez et al. 2010).

3.3 PRODUCTION OF CHITOSAN FROM NATURAL SOURCES

The features of purified chitin, such as degrees of deacetylation, polydispersity index, molecular weight, and purity are regulated by the conditions and parameters of the extraction process. Extraction of chitosan is broadly divided into two categories:

1. Chemical extraction

2. Biological or enzymatic extraction

3.3.1 Chemical Extraction Method

The crustacean shells procured from different sources are washed and dried. The extraction of chitosan involves three steps: (i) demineralization, (ii) deproteinization, and (iii) deacetylation. For the purpose of demineralization, dilute hydrochloric acid (HCl), sulphuric acid (H_2SO_4), acetic acid (CH_3COOH), nitric acid (HNO_3), and formic acid (HCOOH), out of which HCl is considered as the better agent (Percot, Viton, and Domard 2003). This is carried out to remove the calcium carbonate content, which is the major constituent of crustacean exoskeletons. The material obtained after acid treatment is then filtered and washed with distilled water to neutralize. The material is further dried overnight in an oven.

Deproteinization is carried out to remove the protein content from the product and involves the cleavage of the chitin protein bonds, i.e., use of chemicals for depolymerization of the biopolymer. Sodium hydroxide (NaOH) with the concentration ranging from 0.125M to 5.0 M is the most preferred reagent applicable in a wide range of deproteinization reaction conditions (temperature, pressure, and duration). In addition to deproteinization, NaOH facilitates polymer hydrolysis, partial deacetylation of chitin, and reduction in molecular weight (Benhabiles et al. 2013). These steps are followed by decolorization to remove such pigments as β-carotene and astaxanthin by the application of various inorganic and organic solvents, including acetones, hydrogen peroxide and sodium hypochlorite (Dhillon et al. 2013).

3.3.2 Biological Extraction

The evolution and expansion of biological extraction techniques are focused on the "green chemistry" concept. Khanafari et al. published a comparative study between isolation of chitin via chemical and biological routes (Khanafari, Marandi, and Sanatei 2008). The results implied that the biological method was exceedingly better than the chemical extraction method; not only was the procedure was eco-friendly, but the structural integrity of the procured chitin was preserved. In biological extraction, the deproteinization of the crustacean shells is attained using proteolytic microbes (Dhillon et al. 2013). The two methods used for the biological extraction of chitin are microbial fermentation using *Lactobacillus* (Prameela et al. 2010), *Pseudomonas* sp. (Ghorbel-Bellaaj et al. 2011), *Bacillus* sp. (Ghorbel-Bellaaj et al. 2012; Sini et al. 2007), and *Aspergillus* sp. (Mahmoud, Ghaly, and Arab 2007), as well as enzymatic deproteination using pancreatin, devolyase, pepsin, papain, alcalase, and trypsin (Arbia et al. 2013).

The biological/enzymatic deacetylation of chitin to chitosan is carried out in the presence of the enzyme chitin deacetylase. Chitin deacetylase belongs to the carbohydrate esterase family that is used in hydrolysis of the acetamido group of the *N*-acetylglucosamine units of chitin to produce acetic acid and glucosamine units. The deacetylase can also be extracted from organisms such as bacteria (*V. cholera*), insects (*Helicoverpa armigera, Helicoverpa armigera, Drosophila melanogaster*), and fungi (*C. Lindemuthianum, F. Velutipes, M. Racemosus, A. Niger*). In order to increase the efficiency of enzymatic deacetylation, such physical treatments as grinding, sonication, derivatization, and heating occur before deacetylation (Zhao, Park, and Muzzarelli 2010).

3.4 CHEMICAL MODIFICATIONS OF CHITOSAN

Chemical modification of chitosan has gained considerable attention because of their tuneable properties. The ease of chemical modification of chitin and chitosan is mainly attributed to the presence of reactive functional moieties on the polysaccharide backbone. The amino functionality qualifies chitosan for reactions such as quaternization, grafting, alkylation, and reaction with carbonyl compounds. The presence of the hydroxyl group allows chitosan to form hydrogen bonding with polar atoms, grafting, crosslinking, and some chemical reactions such as *O*-acetylation.

(Croisier and Jérôme 2013). These reactive functional groups enable chitosan and chitin to be conveniently modified into various forms such as films, scaffolds, membranes, gels, microparticles, nanoparticles, nanofibrils, beads, sponges, and nanofibers (Alves and Mano 2008). Chitin and chitosan can be easily modified using several chemical modification techniques, namely, sulphonation, azylation, nitration, methylation, N-succinylation, hydroxylation, acylation, thiolation, phosphorylation, xanthation, and graft copolymerization. Among the mentioned means of chemical modification, graft copolymerization is the most investigated and effective technique for a plethora of molecular forms (Shukla et al. 2013).

3.5 FABRICATION OF CHITOSAN BIONANOCOMPOSITES

Chitosan has a plethora of applications in the food packaging industry. Several reinforcements like natural extracts, nanofillers, and polymers are incorporated in the chitosan matrix in order to improve the properties required for packaging applications. Souza et al. (2017) explored the influence of various types of water and oil based natural antioxidants such as kenaf, ginger, rosemary, green and black tea and sage leaves, and five essential oils derived from thyme, rosemary, ginger, tea tree, and sage, respectively on the chitosan films. The addition of the supplements influenced the transmittance of the chitosan films (nearly 15% to 80% reduction compared to the control films). An average increase of 20 to 27 MPa was observed in the tensile strength of the manufactured films (Souza et al. 2017). Turmeric-combined chitosan films generated using a crosslinking agent such as sodium sulfate exhibit excellent antibacterial action against gram-negative and gram-positive bacteria, along with enhanced tensile strength. Incorporation of turmeric did not affect the thermal stability and vapor permeability of the synthesized chitosan-based films. However, the light transparent barrier and stiffness of the film were significantly enhanced. Hence, turmeric-composited chitosan films can be effectively used as packaging material (Kalaycıoğlu et al. 2017).

Nanoengineering of chitosan films is accomplished by rein-forcing various nanostructured fillers such as nanoclay, nano-carbon structures, inert metallic nanoparticles, and functional particles (Ghaani et al. 2016). For instance, reinforcement of montmorillonite clay nanoparticles results in enhancement of oxygen permeability (Priolo, Gamboa, and Grunlan 2009) and mechanical attributes (Wang et al. 2005) of the composite. The incorporation of silver nanoparticles (NPs) in the chitosan matrix enhances the antimicrobial, gas barrier, and mechanical characteristics (Wei et al. 2009). Furthermore, embedding zein protein treated fibrous clays, such as palygorskite and sepiolite, in the polysaccharide network results in significant improve-ment of oxygen and UV light barrier properties, along with water resistance and barrier attributes (Alcântara et al. 2016). Carbon nanotubes (CNTs) (Marroquin, Rhee, and Park 2013), magnetite, and graphene oxide (GO) (Wang et al. 2010) are popularly being utilized as reinforcing agents (Wang, Qian, and Ding 2018).

Liang et al. used zein protein for the encapsulation of a polyphenol, the epigallocatechin gallate (EGCG) that is iso-lated from green tea by blending it with chitosan NPs via electrostatic interaction under mild reaction conditions (Liang et al. 2017). The resulting system established an outstanding means for delivering EGCG in a sustained way for a longer period. Cui et al. reported the use of liposomes for encapsulat-ing pharmaceuticals and other active materials (Cui, Yuan, and Lin 2017). Approximately 57.66±0.12% of bacteriophage was incorporated into chitosan film to maintain the stability of the encapsulated liposomes. The obtained composite film exhibited higher antibacterial activity and effective protection of beef without compromising its quality and sensing properties.

The presence of amino groups on the structural backbone of chitosan enables it to undergo chemical alterations. Interac-tion of chitosan with other polymers involves structural mod-ifications that lead to considerable changes in its chemical, biological, and physical attributes. Exploiting this feature, many studies have been conducted in recent years regarding the blending of polymers with chitosan to attain chitosan-based bioplastics with required properties. Deng et al. reported the cellulose nanofiber (CNF) with methylcellulose, carboxymethylcellulose, and chitosan. This study revealed that the change in physicochemical properties is influenced by the nature of the specific polymer. Association of CNF with chitosan led to considerable lowering of hydrophilicity of the films which reflects much higher compatibility of chitosan with CNF in comparison to other polymers. The fabricated CS-CNF films were further investigated for packaging of some beef patties. The results showed effective prevention of moisture migration within the patties, thus lowering the loss of moistness and layer adhesion. These types of films can conveniently replace petroleum-based packaging materials of synthetic origin (Deng, Jung, and Zhao 2017). Both CNF and CS-CNF have recognizable film-forming features. A 50:50 composition of the blend yielded the best result. This composition exhibited remarkable characteristics such as higher stress, excellent mechanical attributes, and high elongation at break. These characters were further enhanced by increasing chitosan content up to 50% and then decreasing it. The planar surface and homogeneity at cross sections were confirmed by SEM. The compatibility was measured using spectral and x-ray analyses. However, antibacterial activity of the films was not reported. An active, edible, and biodegradable film is formulated by combining different percentages of chitosan with cornstarch in solution via casting. Mechanical and spectral analyses of the resultant blend showed the hydrogen bonding between the amino and hydroxyl groups of both polymers. X-ray studies explained the minimization of crystallinity. The incorporation of chitosan led to the enhancement of tensile strength, solubility, etc., indicating flexibility of the bonds. The vapor permeability also increased, as did chitosan content in the blend (Ren et al. 2017).

3.6 APPLICATIONS OF CHITOSAN IN THE FOOD INDUSTRY

The antimicrobial property of chitosan is a very important characteristic of a substance to determine its fitness for food preservation and packaging applications. This property of chitosan and its derivatives has been widely explored for their use as food preservatives that decelerate the deterioration of food particles caused by microorganisms (Barikani et al. 2014). El-Diasty et al. investigated the result of incorporation of chitosan in cheese to study its effect on the microbial spoilage of cheese (El-Diasty, Eleiwa, and Aideia 2012). It was found that the shelf life of chitosan-infused cheese was considerably elongated by inhibition of mold and yeast growth. The antimicrobial properties of chitosan are also harnessed for packaging and antimicrobial coatings of edible substances. Vegetables, fish, fruits, grains, and other foods can be laminated with chitosan biofilms that act as a layer of protection from microbial activity, thus preserving the nutritional integrity of food (Aranaz et al. 2009). Chitosan biocomposite films with antimicrobial fillers such as antibiotics and chelating agents, as well as plant products such as cinnamaldehyde and thymol, bacteriocins can be utilized to increase the shelf life of food items by inhibiting the growth of microbes and thereby reduce spoilage (Dutta, Tripathi, and Dutta 2012).

Modified chitosan, such as diethoxyphosphorylpolyaminoethyl chitosan (DPECS) and hydroxypropyl chitosan (HPCS), exhibit antibacterial properties very similar to chitosan. Deng et al. inactivated nanofibrous matrix with HPCS via electrospinning, along with addition of organic rectorite to improve its antibacterial properties (Deng et al. 2013). The synthesized matrices exhibited excellent antibacterial action against *S. aureus* and *E. coli*. Fan et al. studied the antifungal activity of modified diethoxyphosphorylpolyaminoethyl chitosan against *B. cinera*, *F. solani*, and *P. capsici* (Fan et al. 2018). The antifungal activity is credited to the presence of phosphoryl and multi-aminoethyl groups. The infusion of hydrophobic groups on to the glycosidic groups can elevate the total hydrophobicity of the chitosan. This in turn enhances the surface activity that plays a major role in loading and sustained release of hydrophibic antimicrobial agents or drugs, which is of utmost importance in packaging (Tan et al. 2015). Furthermore, these chitosan films should have the potential to protect the packaged food items from UV radiations; prevent decay due to oxidation and growth of pathogens; and maintain the handling features and mechanical integrity of the packaged items. All the previously discussed thermal, mechanical, barrier, antimicrobial, antioxidant, and UV shielding features help to elongate the shelf life of food.

ACKNOWLEDGMENTS

Authors express their thanks to TEQIP-III for providing scholarship to Ms. A. Biswal to pursue her doctorate degree.

REFERENCES

Alcântara, Ana CS, Margarita Darder, Pilar Aranda, André Ayral, and Eduardo Ruiz-Hitzky. 2016. "Bionanocomposites based on polysaccharides and fibrous clays for packaging applications." *Journal of Applied Polymer Science* no. 133:2.

Alves, Natalia M, and Joaifo Filipe Mano. 2008. "Chitosan derivatives obtained by chemical modifications for biomedical and environmental applications." *International Journal of Biological Macromolecules* no. 43 (5):401–414.

Anaya, P, G Cárdenas, Vladimir Lavayen, A García, and Colm O'Dwyer. 2013. "Chitosan gel film bandages: Correlating structure, composition, and antimicrobial properties." *Journal of Applied Polymer Science* no. 128 (6):3939–3948.

Aranaz, Inmaculada, Marian Mengíbar, Ruth Harris, Inés Paños, Beatriz Miralles, Niuris Acosta, Gemma Galed, and Ángeles Heras. 2009. "Functional characterization of chitin and chitosan." *Current Chemical Biology* no. 3 (2):203–230.

Arbia, Wassila, Leila Arbia, Lydia Adour, and Abdeltif Amrane. 2013. "Chitin extraction from crustacean shells using biological methods—A review." *Food Technology and Biotechnology* no. 51 (1):12–25.

Barikani, Mehdi, Erfan Oliaei, Hadi Seddiqi, and Hengameh Honarkar. 2014. "Preparation and application of chitin and its derivatives: A review." *Iranian Polymer Journal* no. 23 (4):307–326.

Benhabiles, MS, Nadia Abdi, Nadjib Drouiche, Hakim Lounici, Andre Pauss, Mathews FA Goosen, and Nabil Mameri. 2013. "Protein recovery by ultrafiltration during isolation of chitin from shrimp shells *Parapenaeus longirostris.*" *Food Hydrocolloids* no. 32 (1):28–34.

Croisier, Florence, and Christine Jérôme. 2013. "Chitosan-based biomaterials for tissue engineering." *European Polymer Journal* no. 49 (4):780–792.

Cui, Haiying, Lu Yuan, and Lin Lin. 2017. "Novel chitosan film embedded with liposome-encapsulated phage for biocontrol of *Escherichia coli* O157: H7 in beef." *Carbohydrate Polymers* no. 177:156–164.

Demitri, Christian, Vincenzo Maria De Benedictis, Marta Madaghiele, Carola Esposito Corcione, and Alfonso Maffezzoli. 2016. "Nanostructured active chitosan-based films for food packaging applications: Effect of graphene stacks on mechanical properties." *Measurement* no. 90:418–423.

Deng, Hongbing, Penghua Lin, Wei Li, Shangjing Xin, Xue Zhou, and Jianhong Yang. 2013. "Hydroxypropyl chitosan/organic rectorite-based nanofibrous mats with intercalated structure for bacterial inhibition." *Journal of Biomaterials Science, Polymer Edition* no. 24 (4):485–496.

Deng, Zilong, Jooyeoun Jung, and Yanyun Zhao. 2017. "Development, characterization, and validation of chitosan-adsorbed cellulose nanofiber (CNF) films as water resistant and antibacterial food contact packaging." *LWT-Food Science and Technology* no. 83:132–140.

Dhillon, Gurpreet Singh, Surinder Kaur, Satinder Kaur Brar, and Mausam Verma. 2013. "Green synthesis approach: Extraction of chitosan from fungus mycelia." *Critical Reviews in Biotechnology* no. 33 (4):379–403.

Dutta, Joydeep, S Tripathi, and Pradeep Kumar Dutta. 2012. "Progress in antimicrobial activities of chitin, chitosan, and its oligosaccharides: A systematic study of needs for food applications." *Food Science and Technology International* no. 18 (1):3–34.

El-Diasty, Eman Mahamoud, Nesreen Z Eleiwa, and Hoda AM Aideia. 2012. "Using of chitosan as antifungal agent in Kariesh cheese." *New York Science Journal* no. 5 (9):5–10.

Fan, Zhaoqian, Yukun Qin, Song Liu, Ronge Xing, Huahua Yu, Xiaolin Chen, Kecheng Li, and Pengcheng Li. 2018. "Synthesis, characterization, and antifungal evaluation of diethoxyphosphoryl polyaminoethyl chitosan derivatives." *Carbohydrate Polymers* no. 190:1–11.

Ferraro, Vincenza, Isabel B Cruz, Ruben Ferreira Jorge, F Xavier Malcata, Manuela E Pintado, and Paula ML Castro. 2010. "Valorisation of natural extracts from marine source focused on marine by-products: A review." *Food Research International* no. 43 (9):2221–2233.

Ghaani, Masoud, Carlo A Cozzolino, Giulia Castelli, and Stefano Farris. 2016. "An overview of the intelligent packaging technologies in the food sector." *Trends in Food Science and Technology* no. 51:1–11.

Ghorbel-Bellaaj, Olfa, Kemel Jellouli, Islem Younes, Laila Manni, Mohamed Ouled Salem, and Moncef Nasri. 2011. "A solvent-stable metalloprotease produced by *Pseudomonas aeruginosa* A2 grown on shrimp shell waste and its application in chitin extraction." *Applied Biochemistry and Biotechnology* no. 164 (4):410–425.

Ghorbel-Bellaaj, Olfa, Islem Younes, Hana Maâlej, Sawssen Hajji, and Moncef Nasri. 2012. "Chitin extraction from shrimp shell waste using *Bacillus* bacteria." *International Journal of Biological Macromolecules* no. 51 (5):1196–1201.

Kalaycıoğlu, Zeynep, Emrah Torlak, Gülşen Akın-Evingür, İlhan Özen, and F. Bedia Erim. 2017. "Antimicrobial and physical properties of chitosan films incorporated with turmeric extract." *International Journal of Biological Macromolecules* no. 101:882–888.

Khanafari, A, Reza Marandi, and Sh Sanatei. 2008. "Recovery of chitin and chitosan from shrimp waste by chemical and microbial methods." *Journal of Environmental Health Science and Engineering* no. 5 (1):1–24.

Liang, Jin, Hua Yan, Jiuya Zhang, Wenzhong Dai, Xueling Gao, Yibin Zhou, Xiaochun Wan, and Pradeep Puligundla. 2017. "Preparation and characterization of antioxidant edible chitosan films incorporated with epigallocatechin gallate nanocapsules." *Carbohydrate Polymers* no. 171:300–306.

Mahmoud, Nesreen S, Abdel E Ghaly, and Fereshteh Arab. 2007. "Unconventional approach for demineralization of deproteinized crustacean shells for chitin production." *American Journal of Biochemistry and Biotechnology* no. 3 (1):1–9.

Marroquin, Jason Baxter, Kyong Yop Rhee, and Soo Jin Park. 2013. "Chitosan nanocomposite films: Enhanced electrical conductivity, thermal stability, and mechanical properties." *Carbohydrate Polymers* no. 92 (2):1783–1791.

Percot, Aline, Christophe Viton, and Alain Domard. 2003. "Characterization of shrimp shell deproteinization." *Biomacromolecules* no. 4 (5):1380–1385.

Prameela, Kandra, Challa Murali Mohan, PV Smitha, and KPJ Hemalatha. 2010. "Bioremediation of shrimp biowaste by using natural probiotic for chitin and carotenoid production: An alternative method to hazardous chemical method." *International Journal of Applied Biology and Pharmaceutical Technology* no. 1 (3):903–910.

Priolo, Morgan A, Daniel Gamboa, and Jaime C Grunlan. 2009. "Transparent clay–polymer nano brick wall assemblies with tailorable oxygen barrier." *ACS Applied Materials and Interfaces* no. 2 (1):312–320.

Ramírez, Miguel A, Aida T Rodríguez, Luis Alfonso, and Carlos Peniche. 2010. "Chitin and its derivatives as biopolymers with potential agricultural applications." *Biotecnología Aplicada* no. 27 (4):270–276.

Ren, Lili, Xiaoxia Yan, Jiang Zhou, Jin Tong, and Xingguang Su. 2017. "Influence of chitosan concentration on mechanical and barrier properties of corn starch/chitosan films." *International Journal of Biological Macromolecules* no. 105:1636–1643.

Revathi, Masilamani, Ramachandran Saravanan, and Annaian Shanmugam. 2012. "Production and characterization of chitinase from *Vibrio* species, a head waste of shrimp *Metapenaeus dobsonii* (Miers, 1878) and chitin of *Sepiella inermis* (Orbigny, 1848)." *Advances in Bioscience and Biotechnology* no. 3 (04):392.

Shukla, Sudheesh K, Ajay K Mishra, Omotayo A Arotiba, and Bhekie B Mamba. 2013. "Chitosan-based nanomaterials: A state-of-the-art review." *International Journal of Biological Macromolecules* no. 59:46–58.

Sini, Theruvathil Karunakaran, Sethumadhavan Santhosh, and Paruthapara Thomas Mathew. 2007. "Study on the production of chitin and chitosan from shrimp shell by using *Bacillus subtilis* fermentation." *Carbohydrate Research* no. 342 (16):2423–2429.

Souza, Claudiana P, Bianca C Almeida, Rita R Colwell, and Irma NG Rivera. 2011. "The importance of chitin in the marine environment." *Marine Biotechnology* no. 13 (5):823.

Souza, Victor Gomes Lauriano, Ana Luisa Fernando, João Ricardo Afonso Pires, Patricia Freitas Rodrigues, Andreia AS Lopes, and Francisco M Braz Fernandes. 2017. "Physical properties of chitosan films incorporated with natural antioxidants." *Industrial Crops and Products* no. 107:565–572.

Tan, Chen, Yating Zhang, Shabbar Abbas, Biao Feng, Xiaoming Zhang, Wenshui Xia, and Shuqin Xia. 2015. "Biopolymer–lipid bilayer interaction modulates the physical properties of liposomes: Mechanism and structure." *Journal of Agricultural and Food Chemistry* no. 63 (32):7277–7285.

Venkatesan, Jayachandran, and Se-Kwon Kim. 2010. "Chitosan composites for bone tissue engineering—An overview." *Marine Drugs* no. 8 (8):2252–2266.

Wang, Hongxia, Jun Qian, and Fuyuan Ding. 2018. "Emerging chitosan-based films for food packaging applications." *Journal of Agricultural and Food Chemistry* no. 66 (2):395–413.

Wang, Shaofeng, Lu Shen, Yuejin Tong, Ling Chen, Inyee Phang, PQ Lim, and Tianxi Liu. 2005. "Biopolymer chitosan/montmorillonite nanocomposites: Preparation and characterization." *Polymer Degradation and Stability* no. 90 (1):123–131.

Wang, Xiluan, Hua Bai, Zhiyi Yao, Anran Liu, and Gaoquan Shi. 2010. "Electrically conductive and mechanically strong biomimetic chitosan/reduced graphene oxide composite films." *Journal of Materials Chemistry* no. 20 (41):9032–9036.

Wei, Dongwei, Wuyong Sun, Weiping Qian, Yongzhong Ye, and Xiaoyuan Ma. 2009. "The synthesis of chitosan-based silver nanoparticles and their antibacterial activity." *Carbohydrate Research* no. 344 (17):2375–2382.

Zhao, Yong, Ro-Dong Park, and Riccardo AA Muzzarelli. 2010. "Chitin deacetylases: Properties and applications." *Marine Drugs* no. 8 (1):24–46.

CHAPTER 4

Thermal and Mechanical Studies for Chitin and Chitosan Bionanocomposites

Wen Shyang Chow

School of Materials and Mineral Resources Engineering,
Engineering Campus
Universiti Sains Malaysia
Nibong Tebal, Malaysia

4.1 INTRODUCTION

The use of bionanocomposites is an attractive approach for food packaging materials attributed to their tuneable properties through the combination of biopolymers and nanofillers, biodegradability, and sustainability (Darder et al. 2007). Chitin is the major component of exoskeletons of arthropods and cell walls in fungi. The insolubility of chitin in common solvents is due to its high crystallinity and hydrogen bonding among hydroxyl, carbonyl, and acetamide functional groups. Chitin is often converted to chitosan to increase its usability. Chitosan possesses good biodegradability, biocompatibility, film forming-ability, non-toxicity, and antimicrobial behavior. However, poor mechanical properties and barrier properties have limited its utilization in packaging applications (Bensalem et al. 2017). Some of the feasible strategies for modifying the thermal, mechanical, and barrier properties of chitosan are polymer blending and bionanocomposite technologies. Nanomaterials can be used to tailor the properties of food packaging in order to improve the quality and expand the shelf life of food products (Reig et al. 2014). This chapter discusses the thermal and mechanical properties of chitin and chitosan bionanocomposites. Strategies used to enhance the thermal and mechanical properties of bionanocomposites are explored, for example, use of surface functionalized nanofillers, hybridization of nanofillers, and fabrication techniques.

4.2 THERMAL PROPERTIES

The thermal characteristics of bionanocomposites can be measured using various thermal analyzers, e.g., dynamic mechanical analyzer (DMA), thermogravimetric analyzer (TGA), and differential scanning calorimetry (DSC). Some of the thermal characteristics, including melting temperature (T_m), glass transition temperature (T_g), crystallization temperature (T_c), degree of crystallinity (χ_c), decomposition temperature (T_d), storage modulus (E'), loss modulus (E"), and tan δ, are important to be used as a reference point in controlling the processing and properties of bionanocomposites.

Al-Sagheer and Muslim (2010) fabricated chitosan-silica hybrid films by sol-gel method using tetraethoxysilane (TEOS) as a precursor. The storage modulus (E') of the chitosan increases with the increase in silica loading. Figure 4.1 shows the DMA curves of the chitosan-silica nanocomposites. The E' (at 50°C) of pure chitosan (3.67 GPa) has been increased to 6.70 GPa in the presence of SiO_2 nanoparticles. Kasirga et al. (2012) reported that the T_{50} (the temperature recorded at 50% weight loss from TGA) of the chitosan/montmorillonite nanocomposites (with montmorillonite loading = 5%) is higher than that of chitosan. This is due

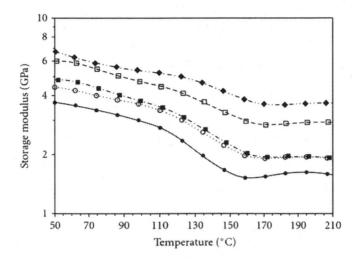

FIGURE 4.1 Temperature Variation of Storage Modulus for CS-Si Nanocomposites at 2 Hz. Note: Silica 0 wt% (black dot symbol), 5 wt% (circle symbol), 10 wt% (black square symbol), 20 wt% (hollow square symbol). (Reprinted with permission from Al-Sagheer F., and Muslim, S. (2010). Thermal and mechanical properties of chitosan/SiO₂ hybrid composites. *Journal of Nanomaterials* 2010: Article ID 490679, 1–7, Hindawi Publishing Corporation).

to two factors: (1) the heat barrier effect of montmorillonite and (2) the interaction of montmorillonite and chitosan, which could delay the thermal decomposition of the nanocomposites.

Swain et al. (2014) prepared chitosan/clay bionanocomposites by solution technique using CuSO₄/glycine chelate complex as a catalyst. From TGA results, the thermal decomposition temperature of chitosan/clay bionanocomposites is higher than that of pure chitosan. This is due to the good interaction between chitosan and clay, as well as to the high thermal stability of clay.

Masheane et al. (2016) studied the effects of functionalized multi-walled carbon nanotube (f-MWCNT) on chitosan-alumina nanocomposites prepared by phase inversion techniques. TGA results showed that the thermal stability of chitosan was improved by the incorporation of alumina and f-MWCNT. Figure 4.2 shows the TGA and DTG curves of chitosan, chitosan-alumina, and chitosan-alumina/f-MWCNTs nanocomposites. Note that an additional weight loss was recorded at 890°C for the chitosan-alumina/f-MWCNTs nanocomposites (Figure 4.2c), which is attributed to the thermal decomposition of f-MWCNT.

According to Cobos et al. (2017), the thermal stability of chitosan was increased by the incorporation of graphene oxide.

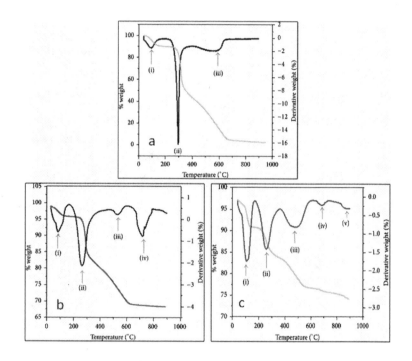

FIGURE 4.2 TGA and TGA Derivative Curves of (A) Chitosan, (B) Chitosan-alumina, and (C) Chitosan-alumina/f-MWCNTs Nanocomposites. (Reprinted with permission from Masheane, M., Nthunya, L., Malinga, S., Nxumalo, E., Barnard, T., Mhlanga, S. 2016. Antimicrobial properties of chitosan-alumina/f-MWCNT nanocomposites. *Journal of Nanotechnology* 2016: Article ID 5,404529, 1–8, Hindawi Publishing Corporation).

The improvement in thermal stability of chitosan may result from the formation of the carbonaceous layer that would protect the underlying polymer from the heat flux. Abbasian and Mahmoodzadeh (2017) prepared antibacterial silver–chitosan-modified bionanocomposites by reversible addition–fragmentation chain transfer (RAFT) polymerization and chemical reduction methods. From TGA results, it can be seen that thermal stability of the silver/chitosan-g-PAA/montmorillonite nanocomposites is higher than that of chitosan-g-PAA copolymer.

By using the DSC method, Rong et al. (2017) reported that the χ_c of chitosan was increased up to 40.6% by the incorporation of cellulose nanowhiskers (10% loading). Salaberría et al. (2018) studied the effects of chitin nanocrystals on the thermal properties of chitosan bionanocomposites. It was found that the free volume at temperatures in the vicinities of α-relaxation was increased, while the E_a of β-relaxation was decreased, in the presence of chitin nanocrystals.

4.3 MECHANICAL PROPERTIES

The durability of the bionanocomposite packaging materials is always related to their mechanical properties, e.g., tensile modulus, tensile strength, and elongation at break. The enhancement in the mechanical properties is mainly attributed to the interfacial interactions among the biopolymer matrices (chitin, chitosan) and nanofillers, as well as the dispersion ability of the nanofiller.

Zhang and Wang (2009) prepared chitosan nanocomposite films filled with Na^+-montmorillonite (MMT) and multiwalled carbon nanotubes (MWCNT) using the solution-evaporation approach. A synergistic effect of MMT and MWCNT on Young's modulus of the chitosan film is achieved when the total loading of MMT and MWCNT is lower than 2.0 wt%. Figure 4.3 shows the schematic structure of the chitosan/MMT/MWCNT nanocomposite film. The $-NH_3^+$ functional groups of chitosans wrapped on the MWCNT's surface could interact with the negatively charged MMT sheets and form the MMT-chitosan-

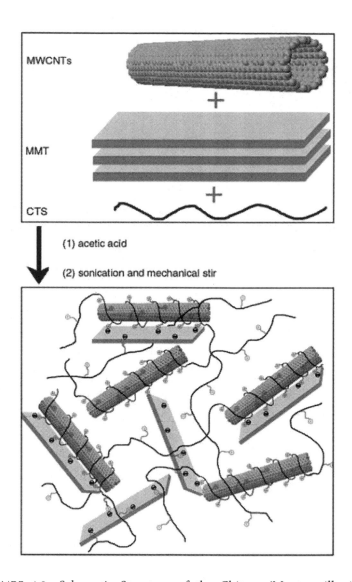

FIGURE 4.3 Schematic Structure of the Chitosan/Montmorillonite/ Multiwalled Carbon Nanotube (CTS/MMT/MWCNTs) Nanocomposite Film. (Reprinted with permission from Zhang, J.P., Wang, A.Q. 2009. Synergistic effects of Na⁺-montmorillonite and multi-walled carbon nanotubes on mechanical properties of chitosan film. *Express Polymer Letters* 3: 302–308, BME-PT).

MWCNT subassembly. This explained why the coaddition of MMT and MWCNT could produce synergistic effects on the mechanical properties of the chitosan nanocomposites.

Hassan et al. (2012) prepared rice straw nanofiller-reinforced chitosan nanocomposites using a solvent casting method. Adding rice straw nanofillers increased the tensile strength of chitosan nanocomposites significantly. This is due to the formation of an interconnected network of nanofibers in the chitosan. Sanuja et al. (2014) prepared a chitosan-magnesium oxide (MgO) nanocomposite film consisting of clove essential oil using a solution casting technique. Figure 4.4 shows the tensile strength and elongation at the break of the chitosan/MgO/clove oil nanocomposites. Incorporating MgO and clove essential oil increased the tensile strength of the chitosan nanocomposite film significantly. This is attributed to the good interaction of MgO and amino functional groups in chitosan.

Soni et al. (2016) prepared TEMPO (2,2,6,6-tetramethylpiperidine-1-oxyl radical)-oxidized cellulose nanofiber (TEMPO-CNF)-reinforced chitosan bionanocomposite films using a casting method. The improvement in the tensile strength of the chitosan films could be due to two factors: (1) the high aspect ratio of TEMPO-CNF (width: 3–20 nm; length: 10–100 nm), and (2) the good interfacial interaction between chitosan and TEMPO-CNF.

Youssef et al. (2016) studied the effects of carboxymethyl cellulose and zinc oxide nanoparticles (ZnO-NP) on the mechanical properties of solution-casted chitosan bionanocomposites. The tensile strength of the chitosan bionanocomposites film was increased from 6.8×10^6 Pa to 12.6×10^6 Pa by the addition of 8 wt% ZnO-NP. This is ascribable to the interaction between ZnO-NP, chitosan (which contains amino groups), and carboxymethyl cellulose (which contains hydroxyl groups).

Gopi et al. (2019) prepared turmeric nanofillers (TNF) from turmeric spent via acid hydrolysis and high pressure homogenization methods and further investigated the influence of TNF on the mechanical properties of chitosan bionanocomposite films.

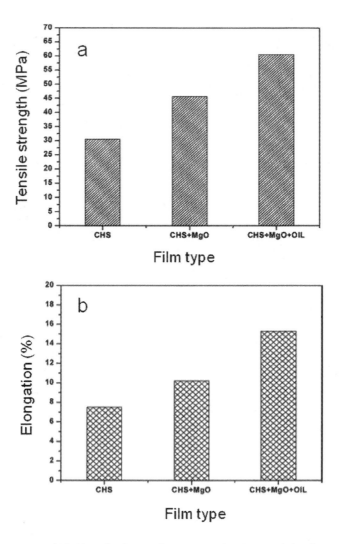

FIGURE 4.4 (A) Tensile Strength versus Film Type; (B) Elongation versus Film Type. Note: CHS: chitosan; CHS+MgO: chitosan/magnesium oxide; CHS+MgO+OIL: chitosan/magnesium oxide/clove oil. (Reprinted with permission from Sanuja, S., Agalya, A., Umapathy, M.J. 2014. Studies on magnesium oxide reinforced chitosan bionanocomposite incorporated with clove oil for active food packaging application. *International Journal of Polymeric Materials and Polymeric Biomaterials* 63: 733–740, Taylor and Francis Group).

The tensile strength and Young's moduli of the bionanocomposites were enhanced significantly as a result of the chitosan–TNF network. Wu et al. (2019) prepared chitosan/ε-polylysine (ε-PL) by in situ self-assembly methods. Sodium tripolyphosphate (TPP) was used as a cross-linking agent. Incorporation of TPP could facilitate the electrostatic interaction and hydrogen bonding between chitosan and ε-PL, and these contributed to the improvement in tensile strength.

Natural nacre possesses and outstanding mechanical properties (e.g., strength, toughness, and light weight) are attributed to the microscale or nanoscale hierarchical structure of chitosan bionanocomposite films, together with specific organic–inorganic interfaces. Using the nacre-like biomimetic structure concept, Yao et al. (2010) fabricated chitosan montmorillonite films by self-assembly methods. The ultimate strength of the artificial nacre-like chitosan-montmorillonite film fabricated by vacuum filtration (76 MPa) is much higher than that of the conventional chitosan/montmorillonite film fabricated by simple mixing (37 MPa).

Enescu et al. (2019) prepared bio-inspired chitosan montmorillonite nanoclays (MMT)/cellulose nanocrystal (CNC) films using a water evaporation–induced self-assembly process. The highest mechanical properties were obtained with an MMT/CNC weight ratio of 1:2. The enhancement of mechanical properties is the result of good interfacial interaction between chitosan, MMT, and CNC. In addition, water evaporation–induced self-assembly is a workable processing technique to produce high performance nacre-like bionanocomposites.

Interfacial interaction, surface modification, and processing techniques play an essential role in controlling thermal and mechanical properties of chitins and chitosan-based bionanocomposites. Functionalization of chitosans or addition of suitable surface modifiers (coupling agent/compatibilizers) are possible approaches to enhancing the properties of the bionanocomposites. In food packaging, while achieving the required mechanical and thermal properties, we should also focus on the barrier properties and antimicrobial behavior of the bionanocomposites. Research focusing on chitosan/nanofiller/antimicrobial agents is a feasible approach for the food packaging industry. Hybridization of multifunctional or smart nanofillers in chitin and chitosan bionanocomposites will be the future trend in developing more sustainable and environmentally friendly food packaging materials.

REFERENCES

Abbasian, M., Mahmoodzadeh, F. 2017. Synthesis of antibacterial silver–chitosan-modified bionanocomposites by RAFT polymerization and chemical reduction methods. *Journal of Elastomers and Plastics* 49: 173–193.

Al-Sagheer, F., Muslim, S. 2010. Thermal and mechanical properties of chitosan/SiO$_2$ hybrid composites. *Journal of Nanomaterials* 2010: Article ID 490679, 1–7.

Bensalem, S., Hamdi, B., Del Confetto, S., Iguer-Ouada, M., Chamayou, A., Balard, H., Calvet, R. 2017. Characterization of chitosan/montmorillonite bionanocomposites byinverse gas chromatography. *Colloids and Surfaces A: Physicochemical and Engineering Aspects* 516: 336–344.

Cobos, M., González, B., Jesús Fernández, M., Dolores Fernández, M. 2017. Chitosan–graphene oxide nanocomposites: Effect of graphene oxide nanosheets and glycerol plasticizer on thermal and mechanical properties. *Journal of Applied Polymer Science* 134 (45092): 1–14.

Darder, M., Aranda, P., Ruiz-Hitzky, E. 2007. Bionanocomposites: A new concept of ecological, bioinspired, and functional hybrid materials. *Advanced Materials* 19: 1309–1319.

Enescu, D., Gardrat, C., Cramail, H., Le Coz, C., Sèbe, G., Coma, V. 2019. Bio-inspired films based on chitosan, nanoclays, and cellulose nanocrystals: Structuring and properties improvement by using water evaporation–induced self-assembly. *Cellulose* 26: 2389–2401.

Gopi, S., Amalraj, A., Jude, S., Thomas, S., Guo, Q.P. 2019. Bionano-composite films based on potato, tapioca starch and chitosan reinforced with cellulose nanofiber isolated from turmeric spent. *Journal of the Taiwan Institute of Chemical Engineers* 96: 664–671.

Hassan, M.L., Fadel, S.M., El-Wakil, N.A., Oksman, K. 2012. Chitosan/ rice straw nanofibers nanocomposites: Preparation, mechanical, and dynamic thermomechanical properties. *Journal of Applied Polymer Science* 125: E216–E222.

Kasirga, Y., Oral, A., Caner, C. 2012. Preparation and characterization of chitosan/montmorillonite-K10 nanocomposites films for food packaging applications. *Polymer Composites* 33: 1874–1882.

Masheane, M., Nthunya, L., Malinga, S., Nxumalo, E., Barnard, T., Mhlanga, S. 2016. Antimicrobial properties of chitosan-alumina/ f-MWCNT nanocomposites. *Journal of Nanotechnology* 2016: Article ID 5404529, 1–8.

Reig, C.S., Lopez, A.D., Ramos, M.H., Cloquell Ballester, V.A. 2014. Nanomaterials: A map for their selection in food packaging applications. *Packaging and Technology Science* 27: 839–866.

Rong, S.Y., Mubarak, N.M., Tanjung, F.A. 2017. Structure–property relationship of cellulose nanowhiskers–reinforced chitosan bio-composite films. *Journal of Environmental Chemical Engineering* 5: 6132–6136.

Salaberría, A.M., Teruel-Juanes, R., Badia, J.D., Fernandes, S.C.M., Sáenz de Juano-Arbona, V., Labidi, J., Riber-Greus, A. 2018. Influence of chitin nanocrystals on the dielectric behaviour and conductivity of chitosan-based bionanocomposites. *Composites Science and Technology* 167: 323–330.

Sanuja, S., Agalya, A., Umapathy, M.J. 2014. Studies on magnesium oxide–reinforced chitosan bionanocomposite incorporated with clove oil for active food packaging application. *International Journal of Polymeric Materials and Polymeric Biomaterials* 63: 733–740.

Soni, B., Barbary Hassan, E., Wes Schilling, M., Mahmoud, B. 2016. Transparent bionanocomposite films based on chitosan- and TEMPO-oxidized cellulose nanofibers with enhanced mechanical and barrier properties. *Carbohydrate Polymers* 151: 779–789.

Swain, S.K., Kisku, S.K., Sahoo, G. 2014. Preparation of thermal resistant gas barrier chitosan nanobiocomposites. *Polymer Composites* 35: 2324–2328.

Wu, C.H., Sun, J.S., Lu, Y.Z., Wu, T.T., Pang, J., Hu, Y.Q. 2019. In situ self-assembly chitosan/ε-polylysine bionanocomposite film with enhanced antimicrobial properties for food packaging. *International Journal of Biological Macromolecules* 132: 385–392.

Yao, H.B., Tan, Z.H., Fang, H.Y., Yu, S.H. 2010. Artificial nacre-like bionanocomposite films from the self-assembly of chitosan-montmorillonite hybrid building blocks. *Angewandte Chemie International Edition* 49: 10127–10131.

Youssef, A.M., EL-Sayed, S.M., EL-Sayed, H.S., Salama, H.H., Dufresne, A. 2016. Enhancement of Egyptian soft white cheese shelf life using a novel chitosan/carboxymethyl cellulose/zinc oxide bionanocomposite film. *Carbohydrate Polymers* 151: 9–19.

Zhang, J.P., Wang, A.Q. 2009. Synergistic effects of Na^+-montmorillonite and multi-walled carbon nanotubes on mechanical properties of chitosan film. *Express Polymer Letters* 3: 302–308.

CHAPTER 5

Barrier, Degradation, and Cytotoxicity Studies for Chitin-Chitosan Bionanocomposites

Aseel T. Issa

High Point Clinical Trials Center
High Point, NC, USA

Reza Tahergorabi

Food and Nutritional Sciences Program
North Carolina Agricultural and Technical State University
Greensboro, NC, USA

5.1 INTRODUCTION: BACKGROUND AND DRIVING FORCES

The structure of chitin contains an amino compound comprised of two monomeric units known as N-acetylglucosamine (GlcNAc) and N-glucosamine (GlcN), randomly distributed throughout the chain, depending on the biopolymer synthesis method used. The ratio of N-acetyl-glucosamine to N-glucosamine structural units determines whether the polymer is chitin or chitosan, with chitin having more than 50% N-acetyl-glucosamine within its structure, while chitosan contains more than 50% N-glucosamine units (Khor and Lim 2003). Chitosan is obtained through the deacetylation of chitin, which is found abundantly in the exoskeletons of a variety of shellfish/invertebrates and cell walls of fungi (Struszczyk 2002).

The degree of deacetylation (DD) determines such properties of the biopolymer as solubility, viscosity, flexibility, and polymer conformation. Chitin has a lower DD forming a crystalline structure that makes it less flexible, insoluble in water and acid, and rigid due to strong hydrogen bonds (Rameshthangam, Solairaj, Arunachalam, and Ramasamy 2018).

The food packaging industry has employed the use of natural polymers based on polysaccharides and protein as a result of their bioactivity, water resistance, permeability, biodegradability, and nontoxicity. However, due to the high cost and high degradation rate of biopolymers, food scientists have employed the use of nanoparticles to fabricate bionanocomposites that may be specifically designed for a particular food product (Vasile 2018). Infusing nanoparticles into bionanocomposites enhances the mechanical, biochemical, and barrier properties by altering the surface morphology of the material to become hydrophilic or hydrophobic, depending on the food packaging requirements. Bionanocomposites incorporated with nanoparticles may also have longer gas infusion paths to increase gas barrier properties and biosensing properties that maintain the flavor, color, texture, and stability of the product (Vasile 2018).

The popularity of chitosan biopolymers in the food packaging industry is due to their biocompatibility, flexibility, low cytotoxicity, and biodegradability. Additionally, chitosan nanocomposites have demonstrated ability to limit bacterial and fungal activity, making them suitable as food coatings. Chitosan nanocomposite scaffolds also have tensile strength within their structure, which makes them highly viscous, and, in turn, suitable for food coatings. According to Jeon, Kamil, and Shahidi (2002), chitosan has been used to make packaging films since 1963 when it was discovered that the use of chitosan films could extend shelf life and reduce the browning of fruit that results from such properties as antibacterial and partial permeability to moisture.

When chitosan is the active component of low-density polyethylene (LDPE) infused with sliver nanoparticles, it reduces cytotoxicity levels while delivering silver ions to the packaged food. In the following sections, studies related to barrier properties, biodegradation, and cytotoxicity of chitin-chitosan bionanocomposites are succinctly reviewed.

5.2 BARRIER PROPERTIES OF CHITIN-CHITOSAN BIONANOCOMPOSITES

Food packaging technology has focused its research on migration and permeability of water vapor, gases, or components within a product or from the wrapping substance to packaged food (Vasile 2018). For instance, fruits and vegetables may require packaging with higher permeability to gases because the freshness and shelf life of such products depend on the regular supply of oxygen to the cells. On the other hand, for products such as carbonated drinks, the containers require high gas barriers to prevent the escape of CO_2 while preventing the entry of oxygen to oxidize the carbonated beverages. Permeability and exchange of ions present a challenge to food packaging scientists due to the extreme changes in atmospheric humidity conditions outside of a controlled experiment (Vasile 2018). Natural polymers obtained from polysaccharides such as chitosan are known to have less flexible swelling capabilities since they rely heavily on the oxygen transfer rate (OTR) to facilitate gaseous exchange and permeability. Swelling is the capacity to absorb fluid and thus aids in the supply of oxygen and nutrients to the interior regions of the nanoscaffolds and increase surface area for cell adherence. The atmospheric humidity levels that constantly fluctuate

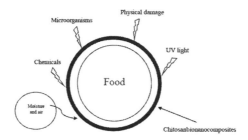

FIGURE 5.1 Barrier Properties of Chitosan Bionanocomposites. It functions as a barrier against moisture and air surrounding the foods. It also protects the food from physical, chemical, and biological deteriorations.

between extreme levels affect the OTR and swelling, which is the capacity to absorb fluid and thus aids in the supply of oxygen and nutrients to the interior regions of the nanoscaffolds (Figure 5.1).

The tensile strength and extensibility of chitosan films are very high. Furthermore, due to their optimum permeability to water vapor and oxygen, these films could be used to extend the shelf life of high-moisture foods (Butler, Vergano, Testin, Bunn, and Wiles 1996; Shahidi and Abuzaytoun 2005).

Azerdo, Attoso, Avena-Bustillos, Munford, Wood, and McHugh (2010) used chitosan-cellulose nanofiber (CSNF) at different concentrations up to 20 (g/100 g in film on dry basis) and glycerol content of up to 30 (g/100 g in film on dry basis) to study mechanical, thermal, and barrier properties of the nanocomposite films. Azerdo et al. (2010) showed that CSNF and glycerol at 15% and 18%, respectively, may enhance mechanical and barrier properties.

Khan, Khan, Salmieri, Le Tien, Riedl, Bouchard, Chauve, Tan, Kamal, and Lacroix (2012) showed that if 5 wt% Chitosan-cellulose nanowhisker (CSNW) were used for fabrication of nanocomposites, it would enhance the tensile strength and water vapor permeability (WVP) of the nanocomposites by 26% and 27%, respectively.

Kerch and Korkhov (2011) indicated that the WVP and mechanical properties of chitosan films have a linear relationship with storage time and molecular weight (M_w) of chitosan and an inverse relationship with storage temperature. They also showed that by incorporating a higher concentration of plasticizer into the film, the mechanical property is enhanced due to less brittleness. Studies also suggest that the colder temperature of the storage room has a positive impact on water vapor uptake by the chitosan films.

5.3 BIODEGRADATION STUDIES OF CHITIN-CHITOSAN BIONANOCOMPOSITES

It has been made possible to biodegrade chitin and chitosan via a nonspecific proteolytic enzyme called lysozyme, in which the output of this degradation is nontoxic for humans (Figure 5.2). There are a few factors contributing to biodegradation of chitin

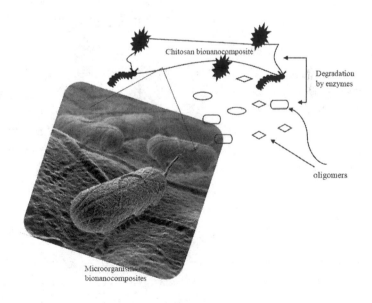

FIGURE 5.2   Biodegradation of Chitosan Bionanocomposites.

and chitosan that include the molecular weight, the pH, the degree of acetylation, and even the method of preparation of chitosan.

5.3.1 In Vitro Studies

Commonly, viscometry and gel permeation chromatography are used in chemical assessment for chitosan degradation. For instance, a 60% loss in viscosity was seen when lysozyme was used to degrade 50% acetylated chitosan after 4h. incubation at pH 5.5. Proteases also can degrade chitosan films. Leucine aminopeptidases have the strongest effect whereby it degrades the films by 38% over 30 days. Kittur, Kumar, and Tharanathan (2003) and Kittur, Kumar, Varadaraj, and Tharanathan (2005) showed that pectinase isozyme from *Aspergillus niger* digests chitosan at low pH.

Chitosan films were also degraded over 4h. and 18 h. by using porcine pancreases. The type of crosslinker has a strong effect on this degradation. For instance, studies have shown that glutaraldehyde is more effective than tripolyphosphate.

5.3.2 In Vivo Studies

In vivo studies related to chitosan degradation is very limited. As a result, the mechanism of action is also unclear. It seems that M_w has a significant effect on the degradation and elimination of chitosan. The liver and kidneys are the probable organs in living organisms in which chitosan is degraded. When it was administered subcutaneously, no change was noticed in rabbits during the time of the experiment (Kean and Thanou 2010).

If chitosan is taken orally, it could be digested and degraded in the intestinal tracts organisms. However, this digestion seems to be species dependent. In hens and broilers, this effect has been noted to be more efficient than in rabbits (Hirano 1991).

In general, DD has a linear relationship with both the rate and extent of chitosan biodegradation in living organisms. Furthermore, enzymatic degradation of chitosan could be affected by N-substitution. This is especially important when systemic administration of new extracts is expected.

5.4 CYTOTOXICITY STUDIES OF CHITIN-CHITOSAN BIONANOCOMPOSITES

Chitosan has been considered as a nontoxic polymer, and few countries—including Japan, Italy, and Finland—have approved it as a food additive. The United States Food and Drug Administration (USFDA) has also approved it for wound healing (Wedmore, McManus, Pusateri, and Holcomb 2006). Chitosan toxicity seems to have a linear relationship with charge density. As the charge density increases the chitosan toxicity is also increased.

5.4.1 In Vitro Studies

Studies on human lymphoblastic leukemia (CCRF-CEM) and human embryonic lung cells (L132) with different M_w and DD of chitosan (M_w <5 kDa, DD 64%; 5–10 kDa, 55.3% DD; and > 10 kDa, 55.3% DD) showed little cytotoxicity (IC_{50} > 1 mg/ml) (Richardson, Kolbe, and Duncan 1999). However, studies on murine melanoma (B16F10) cells with hydrochloride salts of chitosan showed higher levels of toxicity at pH = 5.8 (IC_{50} 0.21 ±0.04 mg/ml) (Carreno-Gomez and Duncan 1997). This result could be due to the pH conditions under which the study was conducted because at pH 5.8 the cell viability is compromised.

In another study, Sanjai, Kothan, Gonil, Saesoo, and Sajomsang (2014) used the gelation method to encapsulate different concentrations of superparamagnetic iron oxide nanoparticles (SPIONPs) inside chitosan-triphosphate (SPIONPs-CS). Low cytotoxicity of SPIONPs CS nanoparticles was reported against skin fibroblast cells at adequate concentrations, and these nanoparticles exhibited excellent stability over prolonged periods of time.

On the other hand, chitosan and its derivatives are toxic against fungi, bacteria, and parasites. This is particularly important when chitin and chitosan nanoparticles are used for food packaging purposes. Jumaa, Furkert, and Müller (2002) showed that chitosan

(87 kDa and 92% DD) in an emulsion system with pH 5–5.3 had a strong antimicrobial effect against *Pseudomonas aeruginosa*, *Staphylococcus aureus*, *Candida albicans*, and *Aspergillus niger*.

5.4.2 In Vivo Studies

Hirano (1991) injected chitosan oligosaccharides with concentration of 7.1–8.6 mg/kg to rabbits for 65 days. As a result, lysozyme activity increased on the first day of injection. However, when the concentration increased to 50 mg/kg, it caused death due to blood aggregation. In contrast, other studies showed no skin or eye irritation in rabbits. In a study on fat chelation, chitosan was administrated at 4.5 g/day in human subjects and no toxicity was reported.

According to most of in vivo studies, chitosan shows minimal toxicity and that is why it has been selected as a safe material in the food packaging industry. However, when new derivatives of chitosan are introduced, safety studies are strongly recommended.

The most challenging issue surrounding the use of bionanocomposites in food packaging is the issue of food safety and toxicity of these materials in case of consumer exposure. It is, however, difficult to determine the safety of nanomaterials due to their small particle size, which makes it difficult to evaluate the chemical and physical characteristics of nanomaterials separately. Thus, food scientists have been navigating safety concerns by determining the estimated risk of public health safety in case of exposure to nanoparticles and other additives in packaging films. Additionally, considerations must be made on the various safety regulation requirements under different jurisdictions in order to design standardized nonmigratory technologies and migratory tests to quantify migrated nanoparticles from packaging systems.

REFERENCES

Azerdo, H., Attoso, L. M., Avena-Bustillos, R., Munford, M., Wood, D. and McHugh, T. H. 2010. Nanocellulose reinforced chitosan composite films as affected by nanofiller loading and plasticizer content. *Journal of Food Science*, 75: 1–7.

Butler, B. L., Vergano, P. J., Testin, R. F., Bunn, J. M. and Wiles, J. L. 1996. Mechanical and barrier properties of edible chitosan films as affected by composition and storage. *Journal of Food Science, 61*: 953–956.

Carreno-Gomez, B. and Duncan, R. 1997. Evaluation of the biological properties of soluble chitosan and chitosan microspheres. *International Journal of Pharmaceutics, 148*: 231–240.

Hirano, S. 1991. Bio-compatibility of chitosan by oral and intravenous administrations. *Polymeric Materials Engineering and Science, 59*: 897–901.

Jeon, Y. J., Kamil, J. Y. and Shahidi, F. 2002. Chitosan as an edible invisible film for quality preservation of herring and Atlantic cod. *Journal of Agricultural and Food Chemistry, 50*: 5167–5178.

Jumaa, M., Furkert, F. H. and Müller, B. W. 2002. A new lipid emulsion formulation with high antimicrobial efficacy using chitosan. *European Journal of Pharmaceutics and Biopharmaceutics, 53*: 115–123.

Kean, T. and Thanou, M. 2010. Biodegradation, biodistribution and toxicity of chitosan. *Advanced Drug Delivery Reviews, 62*: 3–11.

Kerch, G. and Korkhov, V. 2011. Effect of storage time and temperature on structure, mechanical, and barrier properties of chitosan-based films. *European Food Research and Technology, 232*: 17–22.

Khan, A., Khan, R. A., Salmieri, S., Le Tien, C., Riedl, B., Bouchard, J., Chauve, G., Tan, V., Kamal, M. R. and Lacroix, M. 2012. Mechanical and barrier properties of nanocrystalline cellulose-reinforced, chitosan-based nanocomposite films. *Carbohydrate Polymers, 90*: 1601–1608.

Khor, E. and Lim, L. Y. 2003. Implantable applications of chitin and chitosan. *Biomaterials, 24*: 2339–2349.

Kittur, F. S., Kumar, A. B. V. and Tharanathan, R. N. 2003. Low molecular weight chitosans—Preparation by depolymerization with *Aspergillus niger pectinase*, and characterization. *Carbohydrate Research, 338*: 1283–1290.

Kittur, F. S., Kumar, A. B. V., Varadaraj, M. C. and Tharanathan, R. N. 2005. Chitooligosaccharides—Preparation with the aid of pectinase isozyme from Aspergillus niger and their antibacterial activity. *Carbohydrate Research, 340*: 1239–1245.

Rameshthangam, P., Solairaj, D., Arunachalam, G. and Ramasamy, P. 2018. Chitin and chitinases: Biomedical and environmental applications of chitin and its derivatives. *Journal of Enzymes, 1*: 20.

Richardson, S. W., Kolbe, H. J. and Duncan, R. 1999. Potential of low molecular mass chitosan as a DNA delivery system: Biocompatibility, body distribution and ability to complex and protect DNA. *International Journal of Pharmaceutics, 178*: 231–243.

Sanjai, C., Kothan, S. Gonil, P., Saesoo, S. and Sajomsang, W. 2014. Chitosan-triphosphate nanoparticles for encapsulation of superparamagnetic iron oxide as an MRI contrast agent. *Carbohydrate Polymers, 104*: 231–237.

Shahidi, F and Abuzaytoun, R. 2005. Chitin, Chitosan and Co-Products: Chemistry, Production, Applications, and Health Effects. *Advances in Food and Nutrition Research, 49*: 93–135. London: Academic Press.

Struszczyk, M. H. 2002. Chitin and chitosan. Part III. Some aspects of biodegradation and bioactivity. *Polimery, 47*: 619–629.

Vasile, C. 2018. Polymeric nanocomposites and nanocoatings for food packaging: A review. *Materials, 11*: 1834.

Wedmore, I., McManus, J. G., Pusateri, A. E. and Holcomb, J. B. 2006. A special report on the chitosan-based hemostatic dressing: experience in current combat operations. *Journal of Trauma and Acute Care Surgery, 60*: 655–658.

CHAPTER 6

Chitin- and Chitosan-Based Bionanocomposites for Active Packaging

Maria Rapa

University Politehnica of Bucharest
Bucharest, Romania

Cornelia Vasile

Physical Chemistry of Polymers Department
"Petru Poni" Institute of Macromolecular Chemistry
Iași, Romania

6.1 INTRODUCTION

Active packaging represents an efficient way to prolong the life of foodstuffs by using bioactive antibacterial, antioxidant agents, and natural, as well as synthetic, compounds offering various biological functions. It is already known that the right choice of packaging composition offers much fresher foods to consumers. Specific food packaging solutions are required for each specific kind of food (Majid et al. 2018).

Chitins/chitosans (CS) have received considerable interest for food packaging application as components in biocomposites, as nanoparticles (NPs) (CS/chitin NPs) or a combination of various nanofillers and/or vegetable extracts.

Chitosan is a biopolymer derived from renewable resources as chitin, showing interesting physical-chemical properties such as hydrophilicity, solubility in acidic media, adsorption, and bioactivity (antimicrobial/antibacterial and antioxidant, antitumor, and adsorption activities), lipid-lowering film-forming and gelling properties, encapsulation potential, biodegradability, biocompatibility, and nontoxicity (Leceta et al. 2013; Miteluţ et al. 2015). Numerous strategies such as direct solution casting (solvent evaporation), coating, freeze-drying, dipping, layer-by-layer assembly, coprecipitation, complex coacervation, polyelectrolyte complexes, electrospinning, and chemical modification, as well as common processing techniques such as compression molding, thermal molding, melt blending, and twin-screw extrusion were applied to obtain CS/chitin-based films and other materials having valuable multifunctional properties (Hooda et al. 2018; Thomas et al. 2019; Wang et al. 2018). Materials containing CS are found in diverse final products, such as films, fibers/nanofibers, coatings, hydrogels, microspheres, or core/shells structures.

6.2 BIOACTIVE AGENTS FOR FOOD CONTACT COMPOSITES CONTAINING CHITIN/CHITOSAN

6.2.1 Antioxidant Agents

Antioxidant active packaging composite films containing CS have gained an attracting interest in the food field because they are able to delay the oxidative process during the life of foods. β-carotene (Hari et al. 2018), essential oils from perennial herbs (Esmaeili and Asgari 2015; Jahed et al. 2017; Munteanu et al. 2018), lemon essential oils (Hasani et al. 2018), graphene oxide (GO) (Barra et al. 2019), and herbal medicine (Ghelejlu et al. 2016) have been studied as the most promising antioxidant agents for active food packaging together with CS. Mango leaf extract (MLE) incorporated into CS on the cashew nuts preservation for 28 days storage led to an increase with 56% of oxidation resistance, in comparison with a commercial polyamide/polyethylene film (Rambabu et al. 2019).

The incorporation of graphene oxide (GO) hydrothermally reduced (up to 50 wt.% in respect with CS weight) together with caffeic acid in CS matrix significantly increased $ABTS^{•+}$ inhibition with 54–82%, for an incubation period of 8 h. (Barra et al. 2019).

6.2.2 Antimicrobial Agents

Main natural antimicrobials are isolated from microbial (e.g., natamycin, nisin), plant (e.g., essential oil of oregano, basil, cinnamon, thyme, rosemary, and clove), or animal (e.g., lactoferrin, lysozyme) origin, and carboxylic acids (e.g., citric acid, sorbic, propionic) (Cansu Feyzioglu and Tornuk 2016; Wang et al. 2015) are useful in biocomposites containing CS/chitin.

Siripatrawan and Vitchayakitti (2016) reported enhanced antimicrobial activity of CS/propolis extract films (containing 0–20 wt.% propolis extract) against *Escherichia coli*, *Staphylococcus aureus*, *Pseudomonas aeruginosa*, and *Salmonella enteritidis* by using agar diffusion assay.

The strong antimicrobial action of the active CS/ginger essential oil film on the chilled storage of fish steak was evidenced by the determination of the total mesophilic count (Remya et al. 2016) which had a 7 log CFU/g value. Hence, the shelf life of barracuda steak could be increased up to eight days in comparison with reference sample.

In the case of CS/ellagic acid biocomposite films (0.5–5.0 wt.% ellagic acid relative to CS), total inhibition of the *P. aeruginosa* and *S. aureus* pathogenic bacteria was demonstrated by Vilela et al. (2017). Bionanocomposite films included nanofibers based on chitin and CS, which originate from lobster wastes, demonstrated an inhibitory effect (>80%) against *A. Niger* (Salaberria et al. 2015).

6.2.3 Antioxidant and Antimicrobial Agents

There are many active agents acting both as antioxidant and antimicrobial components in CS-composites such as kombucha tea (KT) (Ashrafi et al. 2018), ginger and eugenol essential oils (Bonilla et al. 2018), rosehip seed oil, nanoclay (Butnaru et al. 2019), and herbal medicine based on Carum copticum essential oil (Esmaeili and Asgari 2015).

Extension in the shelf life of ~3 days for minced beef coated with CS/3% kombucha tea extract (KT) active film was demonstrated by Ashrafi et al. (2018). It was based on the retardation of lipid oxidation (proved by a value 0.31 mg/kg malondialdehyde found after 6 days at 4°C) and antimicrobial effect against *E. coli* (increase in microbial activity in the range 5.36–2.11 log CFU/g after 4 days of storage).

6.2.4 Nanoparticles (NPs)

6.2.4.1 Chitosan/Chitin NPs

Nanoparticles (NPs) can be added into chitosan matrix as reinforcement or as active agents. The use of CS NPs to encapsulate bioactive compounds (as essential oils) seems to be an innovative process for prolonging the storage period of food because their degradation during packaging processing is prevented. Haider et al. (2017) encapsulated krill oil (KO) into CS NPs by emulsification followed by electrostatic interaction and used tripolyphosphate as crosslinker. In this way, a delay of KO towards oxidation even after two weeks of storage at 45°C was obtained.

Shao et al. (2018) has prepared a nanoemulsion by ultrasonication based on CS and eugenol in presence of surfactant as an emulsifier. The nanoemulsions exhibited an improved antimicrobial activity against *E.coli*, *S. aureus*, *Salmonella*, and *P. aeruginosa* in comparison with free eugenol, as well as a prolonged shelf life of food formulations. Mohammadi et al. (2015) successfully encapsulated *Cinnamomum zeylanicum* into CS NPs by ionic gelation techniques and applied it as a coating on cucumbers. This coating extended the shelf life of cucumbers by ~6 days at 10 ± 1°C.

Sensory evaluation of chilled pork coated with 527 nm cinnamon essential oil/CS NPs and stored for 12 days at 4°C revealed the maintenance of food quality (Hu et al. 2015).

6.2.4.2 Inorganic NPs as Nanofillers and Antimicrobial Agents

Inorganic antimicrobial agents such as AgNPs (Jafari et al. 2016), ZnO-NPs (Saral Sarojini et al. 2019), and CuO-NPs (Syame et al. 2017) have been used in order to obtain active bionanocomposites films.

Kumar et al. (2018) reported that the packaging of red grapes with hybrid nanomaterial films based on CS, gelatin, and AgNPs 0.1 wt.% prolonged the shelf life of food by 18 days, before mildew occurred.

Youssef et al. (2016) investigated the influence of CS/carboxymethyl cellulose (CMC)/ZnO-NPs bionanocomposite films towards the total yeast and bacterial counts during cheese storage. The investigation showed that the cheese packaged in material containing ZnO-NPs in the range of 2–8% presented an entire white surface after 15 days while the reference samples changed their appearance.

In a separate contribution, Noshirvani et al. (2017) reported an extended shelf life of wheat bread when CMC/CS/oleic acid/ZnO NPs active bionanocomposites were used for packaging. A coating with bionanocomposite film containing 2% ZnO NPs proved the reduced counts of fungi after 22 days of storage at 25°C (Figure 6.1).

Butnaru et al. (2019) found that addition of rosehip seed oil and C30B nanoclay increased antibacterial activity in CS films against *E. coli*. The addition of C30B nanoclay to CS/rosehip seed oil films did not provide significant contribution to change the properties of the films.

Chicken packaged in innovative CS/ε-polylysine nanofiber-based films have shown a significant antibacterial activity against *Salmonella* (Lin et al. 2018). The appearance, flavor, juiciness, color, and overall acceptability values proved the importance of antibacterial packaging on prolonging the shelf life for poultry products (Figure 6.2).

(a)   (b)   (c)

FIGURE 6.1   Visual Appearance of Bread Slices in Contact with *Aspergillus niger*: (A) Control coating after 3 days; (B) CMC/CS-OL/ZnO 0.5% bionanocomposite coating after 11 days; (C) CMC/CS-OL/ZnO 2% bionanocomposite coating after 22 days. The development of yeast and molds is marked with red arrays.

(Adapted from Noshirvani et al. 2017).

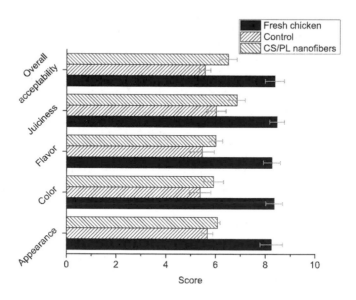

FIGURE 6.2   The Sensory Evaluation of Chicken Samples with or without ε-polylysine/CS Nanofibers Treatments at 4°C for 7 days. Control is CS nanofibers. High score means the negative effect on the sensory analysis.

6.3 CONCLUSIONS AND FUTURE TRENDS

The developed films based on CS/chitin have proved superior antibacterial, mechanical, barrier and sensory properties (Wang et al. 2018). Most of the studies focused on preparation of bioactive food packaging by casting techniques as compared to other polymer processing techniques. The materials designed to extend the shelf life of food must provide antioxidant/antimicrobial activity. A sensory evaluation is necessary to validate the utility of packaging films. The tests and standards for validating the shelf life are still necessary.

In addition to improve antimicrobial and antioxidant properties, researchers should investigate the possibility of upgrading the chitin/CS-based materials for commercial food packaging.

It can be concluded that the exploration of new bioactive packaging biomaterials based on chitin/CS materials has become a new challenge for extending the shelf life of food, while decreasing the environmental problems associated with packaging waste. CS/chitin films have shown much important improvement in bioactive and sensory properties when associated with other film-forming materials and/or natural extracts with antioxidant/antimicrobial/biological functions.

REFERENCES

Ashrafi, A., M. Jokar, and A. M. Nafchi. 2018. Preparation and characterization of biocomposite film based on chitosan and kombucha tea as active food packaging. *Int J Biol Macromol* 108:444–54.

Barra, A., N. M. Ferreira, M. A. Martins, O. Lazar, A. Pantazi, A. A. Jderu, S. M. Newmayer, B. J. Rodrigues, M. Enachescu, P. Ferreira, and C. Nunes. 2019. Eco-friendly preparation of electrically conductive chitosan–reduced graphene oxide flexible bionanocomposites for food packaging and biological applications. *Compos Sci Technol* 173:53–60.

Bonilla, J., T. Poloni, R. V. Lourenço, and P. J. A. Sobral. 2018. Antioxidant potential of eugenol and ginger essential oils with gelatin/chitosan films. *Food Biosci* 23:107–14.

Butnaru, E., E. Stoleru, M. A. Brebu, R. N. Darie-Nita, A. Bargan, and C. Vasile. 2019. Chitosan-based bionanocomposite films prepared by emulsion technique for food preservation. *Materials* 12:1–17.

Cansu Feyzioglu, G., and F. Tornuk. 2016. Development of chitosan nanoparticles loaded with summer savory (*Satureja hortensis* L.) essential oil for antimicrobial and antioxidant delivery applications. *LWT Food Sci Technol* 70:104–10.

Esmaeili, A., and A. Asgari. 2015. In vitro release and biological activities of Carum copticum essential oil (CEO) loaded chitosan nanoparticles. *Int J Biol Macromol* 81:283–90.

Ghelejlu, S. B., M. Esmaiili, and H. Almasi. 2016. Characterization of chitosan–nanoclay bionanocomposite active films containing milk thistle extract. *Int J Biol Macromol* 86:613–21.

Haider, J., H. Majeed, P. A. Williams, W. Safdar, and F. Zhong. 2017. Formation of chitosan nanoparticles to encapsulate krill oil (Euphausia superba) for application as a dietary supplement. *Food Hydrocoll* 63:27–34.

Hari, N., S. Francis, A. G. R. Nair, and A. J. Nair. 2018. Synthesis, characterization and biological evaluation of chitosan film incorporated with β-Carotene loaded starch nanocrystals. *Food Pack Shelf Life* 16:69–76.

Hasani, S., S. M. Ojagh, and M. Ghorban. 2018. Nanoencapsulation of lemon essential oil in Chitosan-Hicap system. Part 1: Study on its physical and structural characteristics. *Int J Biol Macromol* 115:143–51.

Hooda, R., B. Batra, V. Kalra, J. S. Rana, and M. Sharma. 2018. Chitosan-based nanocomposites in food packaging. In *Green and Sustainable Advanced Packaging Materials*, ed. S. Ahmed, 269–84. Singapore: Springer Nature Singapore Pvt. Ltd.

Hu, J., X. Wang, Z. Xiao, and W. Bi. 2015. Effect of chitosan nanoparticles loaded with cinnamon essential oil on the quality of chilled pork. *LWT Food Sci Technol* 63:519–26.

Jafari, H., M. Pirouzifard, M. A. Khaledabad, and H. Almasi. 2016. Effect of chitin nanofiber on the morphological and physical properties of chitosan/silver nanoparticle bionanocomposite films. *Int J Biol Macromol* 92:461–66.

Jahed, E., M. A. Khaledabad, H. Almasi, and R. Hasanzadeh. 2017. Physicochemical properties of *Carum copticum* essential oil loaded chitosan films containing organic nanoreinforcements. *Carbohydr Polym* 164:325–38.

Kumar, S., A. Shukla, P. P. Baul, A. Mitra, and D. Halder. 2018. Biodegradable hybrid nanocomposites of chitosan/gelatin and silver nanoparticles for active food packaging applications. *Food Pack Shelf Life* 16:178–84.

Leceta, I., P. Guerrero, S. Cabezudo, and K. de la Caba. 2013. Environmental assessment of chitosan-based films. *J Clean Prod* 41:312–18.

Lin, L., X. Liao, D. Surendhiran, and H. Cui. 2018. Preparation of ε-polylysine/chitosan nanofibers for food packaging against *Salmonella* on chicken. *Food Pack Shelf Life* 17:134–41.

Majid, I., G. A. Nayik, S. M. Dar, and V. Nanda. 2018. Review article. Novel food packaging technologies: Innovations and future prospective. *J Saudi Soc Agri Sci* 17:454–62.

Miteluţ, A. C., E. E. Tănase, V. I. Popa, and M. E. Popa. 2015. Sustainable alternative for food packaging: chitosan biopolymer – A review. *AgroLife Scientific J* 4:53–61.

Mohammadi, A., M. Hashemi, and S. M. Hosseini. 2015. Chitosan nanoparticles loaded with Cinnamomum zeylanicum essential oil enhance the shelf life of cucumber during cold storage. *Postharvest Biol Technol* 110:203–13.

Munteanu, B. S., L. Sacarescu, A. L. Vasiliu, G. E. Hitruc, G. M. Pricope, M. Sivertsvik, J. T. Rosnes, and C. Vasile. 2018. Antioxidant/antibacterial electrospun nanocoatings applied onto PLA films. *Materials* 11, 1973, 18 pages. doi:10.3390/ma11101973, www.mdpi.com/journal/materials.

Noshirvani, N., B. Ghanbarzadeh, R. R. Mokarram, and M. Hashemi. 2017. Novel active packaging based on carboxymethyl cellulose-chitosan–ZnO NPs nanocomposite for increasing the shelf life of bread. *Food Pack Shelf Life* 11:106–14.

Rambabu, K., G. Bharath, F. Banat, P. L. Show, and H. H. Cocoletzi. 2019. Mango leaf extract–incorporated chitosan antioxidant film for active food packaging. *Int J Biol Macromol* 126:1234–43.

Remya, S., C. O. Mohan, J. Bindu, G. K. Sivaraman, G. Venkateshwarlu, and C. N. Ravishankar. 2016. Effect of chitosan based active packaging film on the keeping quality of chilled stored barracuda fish. *J Food Sci Technol* 53:685–93.

Salaberria, A. M., R. H. Diaz, J. Labidi, and S. C. M. Fernandes. 2015. Role of chitin nanocrystals and nanofibers on physical, mechanical and functional properties in thermoplastic starch films. *Food Hydrocoll* 46:93–102.

Saral Sarojini, K., M. P. Indumathi, and G. R. Rajarajeswari. 2019. Mahua oil-based polyurethane/chitosan/nano ZnO composite films for biodegradable food packaging applications. *Int J Biol Macromol* 124:163–74.

Shao, Y., C. Wu, T. Wu, Y. Li, S. Chen, and Y. Hu. 2018. Eugenol-chitosan nanoemulsions by ultrasound-mediated emulsification: Formulation, characterization and antimicrobial activity. *Carbohydr Polym* 193:144–52. doi:10.1016/j.carbpol.2018.03.101.

Siripatrawan, U., and W. Vitchayakitti. 2016. Improving functional properties of chitosan films as active food packaging by incorporating with propolis. *Food Hydrocoll* 61:695–702.

Syame, S. M., W. S. Mohamed, R. K. Mahmoud, and S. T. Omara. 2017. Synthesis of copper-chitosan nanocomposites and its application in treatment of local pathogenic isolates bacteria. *Orient J Chem* 33. doi:10.13005/ojc/330632.

Thomas, M. S., R. R. Koshy, S. K. Mary, S. Thomas, and L. A. Pothan. 2019. *Starch, Chitin and Chitosan-Based Composites and Nanocomposites. Springer Cham for Springer. Briefs in Molecular Science.* Switzerland AG: Springer Nature. doi:10.1007/978-3-030-03158-9_4.

Vilela, C., R. J. B. Pinto, J. Coelho et al. 2017. Bioactive chitosan/ellagic acid films with UV-light protection for active food packaging. *Food Hydrocoll* 73:120–28.

Wang, H., R. Zhang, H. Zhang et al. 2015. Kinetics and functional effectiveness of nisin loaded antimicrobial packaging film based on chitosan/poly(vinyl alcohol). *Carbohydr Polym* 127:64–71.

Wang, H., J. Qian, and F. Ding. 2018. Emerging chitosan-based films for food packaging applications. *J Agric Food Chem* 17, 66:395–413.

Youssef, A. M., S. M. El-Sayed, H. S. El-Sayed, H. H. Salama, and A. Dufresne. 2016. Enhancement of Egyptian soft white cheese shelf life using a novel chitosan/carboxymethyl cellulose/zinc oxide bionanocomposite film. *Carbohydr Polym* 151:9–19.

CHAPTER 7

A Theoretical Approach to Chitin- and Chitosan-Based Bionanocomposites

K. P. Sajesha

School of Chemical Sciences
Mahatma Gandhi University
Kottayam, India

7.1 INTRODUCTION

Computational chemistry is the application of mathematics to chemistry. It includes the development of algorithms and computer programs to solve chemical programs. Applications of computer programs resulted in the emergence of the small screen laboratory technique called computational chemistry (Jensen 2016). This field helps researchers to save time, effort, and chemicals required for their studies. It also assists in understanding the experimental data, and determining the starting point of a laboratory synthesis, the possible unknown molecule and its geometry, the reaction mechanisms that cannot be predicted by experimental methods, the transition state, the spectra, and physical properties, as well as predicting the nature of interaction of the substrate with enzymes. There are large varieties of program software available for computational studies. Some of the packages are limited to a specific method, whereas the others may include many methods for which choice depends on the nature of the problem under consideration. These programs include molecular modeling (Lewars 2016) programs, molecular mechanics programs, molecular design programs, semi-empirical programs, and quantum mechanical programs. The main tools of computational chemistry include molecular mechanics, which considers the molecule as a ball and stick model, and is used mainly to study large molecules such as steroids, Ab initio theory is based on the Schrödinger equation and electron density determination, and hence can be used to predict the reactivity of systems. Semiempirical methods involve solving the Schrödinger equation along with parameterization with the guidance of experimental values. Density functional theory (DFT) (Young 2004) uses functional theory instead of wave function. Molecular dynamics is based on Newton's laws of motion. Novel molecules, their reactivity, and spectra are most likely studied using ab initio or semiempirical methods, whereas large molecules are studied through molecular mechanics and their active sites by using semiempirical or ab initio methods. Molecular docking addresses the combination of such studies and is most probably applied in the biomedical fields. Computational studies have not only intervened in the field of organic, inorganic, and theoretical chemistry but have also joined their focus with that of polymer chemistry. A simple pictorial representation of the pathways adopted by researchers with the aid of computational chemistry is given in Figure 7.1.

Chitin (Rinaudo 2006), the second most abundant naturally occurring polymer after cellulose contains N-acetyl-glucosamine monomer and its derivative chitosan obtained from enzymatic or chemical deacetylation, has long aroused a high degree of attention within the scientific community. These natural polymers present in the exoskeletons of insects, shellfish—including crabs, lobsters, shrimp, and their derivatives—have vast applications,

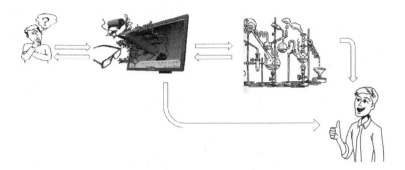

FIGURE 7.1   Research pathways adopting computational chemistry.

including biomedical applications, as wound dressing material, drug delivery vehicles, and tissue engineering implantations. The efficiency of these polymers has been established for widely accepted applications to a large extent because of experimental studies. In addition to these laboratory techniques, the newly emerging areas of computational and theoretical studies has also expanded for further in-depth and thorough studies required to disseminate knowledge on chitin, chitosan, and their derivatives, as well as corresponding nanocomposites, in order to further enable the application studies of these polymers.

7.2 COMPUTATIONAL STUDIES OF CHITIN- AND CHITOSAN-BASED BIOPOLYMERS AS POTENTIAL METAL ION ABSORBERS

Industrialization and modernization in the present era have introduced a substantial quantity of metal ions to the aquatic and terrestrial world that threaten current living organisms. As a result, most of the research is now focused on several biopolymers and their modification to remove these toxic metal ions from the environment. Chitin, chitosan, and some other naturally occurring biopolymers have been established as an effective source for the removal of heavy metals (Popuri et al. 2009) and other pollutants from water.

The nitrogen-containing groups present in the chitin and chitosan unit are important sites for chelation of metal ions in any pH medium forming stable complexes. The binding capacity of metal ions with these polymers is associated with the degree of such factors as deacetylation, surface area, molecular weight, and crystallinity. This absorptive property is applicable for removing toxic radioactive metals and heavy metals from water, recovery of some other metals, nitrogen-based fertilization in the agricultural sector, removal of pesticides, and stabilization of fine clay particles and fuel cells.

Lü et al. (2008) have used Dmol3 calculations based on DFT theory to study the interactions of Cu(II) ions with chitin and chitosan residues. The chelation mechanism was studied with two different models—the bridge model and the pendent model. The bridge model has metal ions connecting several amine groups formed through intermolecular or intramolecular complex formation. The pendant model has metal ions linking an amine group forming a pendantlike structure. Optimizations of the related conformations were done with BLYP/DND methods. It was performed at the generalized gradient approximation (GGA) level using the spin unrestricted and symmetric unrestricted approaches with multiplicity 2. The results favored the bridge model over the pendant model. This can be used to study the mechanism of binding metal ions to chitin and chitosan polymers, which can be further applied for absorption of metal ions.

Molecular dynamic simulation under different conditions has also been used to study chitin- and chitosan-based systems by Franca et al. The simulations were used to characterize the polymer structure in aqueous medium (Franca et al. 2008). The study also had the objective of determining the interactions at molecular level as well as the roles such as solvation, structure, solubility of polymers, influence of ionic strength, and pH. Factors such as intramolecular hydrogen bonding, van der Waals forces, and hydrophobic interactions that indicate the solubility of these polymers were reported. The studies have been done using GROMACS 3.3.2 program, with extensions of GROMOS carbohydrate force field.

Molecular mechanics (MM2) and parameterization model version 3 (PM3) calculations have also been reported by Debbaudt et al. to study the kinetics and metal (lead, mercury, cadmium) absorption on chitin-pectin pellets (Debbaudt et al. 2004). MM2 calculations were used to study the conformational energies and changes that occurred due to the presence of metals. In addition, PM3 calculations were used to determine the enthalpies of formation. The results showed that cadmium absorption is most preferable in chitin pectin pellets.

The intensive absorption of mercury and lead ions from industrial wastes by chitin and chitosan has been studied by Svetlana et al. (2018), thus making it a method for removing these toxic metals from water. Studies have also shown that absorption of mercury is better than absorption of lead, thus providing a method for separating mercury from lead. Functionalized chitosan has also been used for this purpose. Computational approaches using DFT have been used as modern tools providing greater insight for the binding mechanism in chitin, chitosan, and their derivatives with metal ions at molecular levels. These studies have been conducted at the M06-2X/LanL2DZ level using Gaussian09 package.

The DFT studies using Gaussian09 package were also performed by Gomes and his coworkers to study the binding mechanisms (Gomes et al. 2014) of nickel copper and zinc monovalent and divalent ions with chitin and chitosan polymers, which showed that the formation of Cu(II) complexes are more favorable. These studies were also performed using DFT calculations resulting in the studies of bond formation and bond dissociation enthalpies at B3LYP/6-31G** level theory. Figure 7.2 presents the DFT results of the optimized metal (II)-glucosamine $(H_2O)_3$ complexes of nickel, copper, and zinc.

Kim et al. reported the development of regenerated chitin fiber and its nonwoven mat-type separator (Kim et al. 2017). It

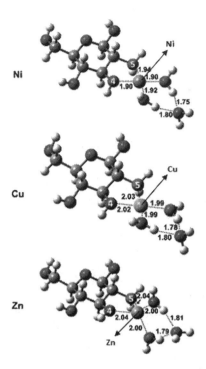

FIGURE 7.2 Optimized structures of metal(II)-glucosamine(H2O)$_3$ complexes(Metals-Ni, Cu, Zn).

suppressed the growth of lithium dendrites in lithium batteries, which upgraded their lithium cycling efficiency. DFT calculations and molecular dynamics simulations were used in these studies, which proved the high affinity of lithium ions onto chitin. These studies confirmed an excellent electrolyte inhibition property in the separators, thus proving that lithium metal batteries are a key requirement of the separators.

Solvent extractions of metal ions such as Cu(II), Ni(II), Co(II), Zn(II), Fe(III), and Pb(II) were performed using lipophilic polysaccharides such as O,O'-decanoyl chitosan and O,O'-decanoyl N,N-didecanoyl chitosan, and O,O'-decanoyl chitin in chloroform (Prasad Dhakal et al. 2005). Molecular modeling was done for chitosan and O,O'-decanoyl chitosan using MOPAC93 to determine the optimum structures and evaluate the related chemical properties. Molecular mechanics calculations were conducted using HyperChem v.5.0 software. These studies predicted the abnormally high selectivity of Cu(II) over Fe(III) with O,O'-decanoyl chitosan that was more effective for studying the associated chemical properties.

The properties of these biopolymers in absorption can also make them applicable in food packaging and biomedical areas that can be widely investigated using computational methods.

In addition to these previously discussed computational studies, there are many other experimental and theoretical studies that support chitin and chitosan-based biopolymers as efficient and effective metal absorbers.

7.3 COMPUTATIONAL STUDIES OF CHITIN- AND CHITOSAN-BASED BIOPOLYMERS IN BIOMEDICAL FIELDS

In the present scenario, drug delivery agents face a serious issue due to such factors as their reduced solubility and absorption and faster degradation, resulting in extensive scientific efforts to overcome this situation. The introduction of nanoparticles over many years has provided a powerful tool for various biomedical applications (Jayakumar et al. 2010). Among these, such biopolymer nanoparticles as cellulose, chitin, and chitosan have proved their efficiency. Chitin and chitosan with nanoparticles form bionanocomposites that provide its new version to the fields of regenerative medicines and drug delivery. Such properties as biocompatibility, bioresourceablity, and biodegradability increased biodistribution, as well as bioavailability, of chitin and chitosan. In addition, the property of protecting drugs from degradation and proper targeting with surface functionalization makes chitin and chitosan most suitable in the biomedical field (Prabaharan 2008) for delivery of drugs and genes, tissue engineering, wound healing, and antimicrobial and anti-viral applications. Chitin and chitosan nanomaterials are also used in bioimaging, and also as biosensors, electrochemical cell sensors, and therapeutic and bioactive implants (Khor and Lim 2003).

Computational methods have also been recently been used to study the biomedical applications of polymer systems. The applications of polymeric nanoparticles to drug delivery systems pose a challenge when experimental techniques are required for their optimization, which is not feasible. This challenge is also associated with the longer amount of time required, and the study of the kinetics and mechanism of drug release, as well as its interaction with the polymer. Therefore, instead of depending on experimental methods for determining suitable carriers in drug delivery, computationally developed techniques have been adopted for studying these systems. In silico methods have also been used recently in designing drugs. Several model systems have been developed that could predict, quantify, and compare the interaction of chitin with drugs and the affinity of binding. These models can be used to study the stability, capacity of loading, and release of drugs from polymeric systems.

Dhanasekaran et al. (2018) have used both in vitro and in silico approaches to study the drug delivering potential and molecular interactions of the drugs with chitin and chitosan-based nanoparticles. Their studies have suggested that chitosan nanoparticles serve as efficient carriers for both curcumin and insulin.

The chitin derivative, chitosan has also been extensively applied in gene therapy. The biocompatible and biodegradable nature of chitosan has made it applicable as a gene carrier. The understanding of the mechanism and kinetics of interactions between DNA strands and chitosan polymers helps in designing chitosan-based drugs and related gene delivery systems.

Ho et al. experimentally modified chitosan (CS) with disulfide and arginine (Arg) forming (CS–SS–Arg), a novel gene carrier (Ho et al. 2011). Molecular dynamics were used to elucidate the self-organized CS–SS–Arg/DNA structures. Figure 7.3 represents the molecular dynamic (MD) simulations studied by Ho and his coworkers. The simulation models of CS, DNA, CS-SS-Arg, and

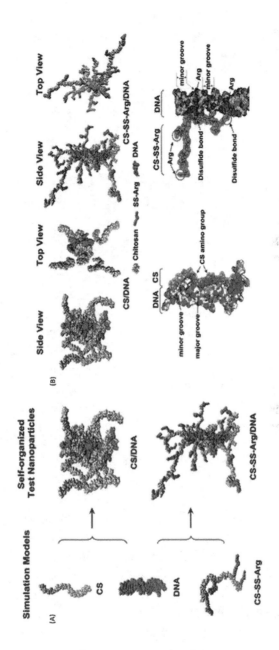

FIGURE 7.3 (A) Chemical structure MD simulation models and their self-organized structures; (B) MD simulation results with the CS/DNA and CS–SS–Arg/DNA self-organized nanocomplexes.

self-organized test structures are given in Figure 7.3A, and the view of the resulting CS/DNA and CS–SS–Arg/DNA from different directions is given in Figure 7.3B.

In addition to those mentioned, many other studies and research associated with the biomedical applications of chitin and chitosan biopolymers are key areas of focus for the scientific community.

7.4 COMPUTATIONAL STUDIES OF CHITIN AND CHITOSAN—PAVING A WAY TO FOOD PACKAGES AND OTHER BIOMATERIALS

The applications of chitin, chitosan, and their modified varieties have also been extended to the engineering of biological materials for sensors, food processing, and packaging for formation of biodegradable films and additive applications for thickening paper. Among these, food packaging is an interesting, newly emerging area as it could overcome many environmental issues such as degradability and disposal. The reason for application of biopolymers in different sectors depends on the nature of interactive, structural, chemical, and thermodynamic properties. However, computational data regarding the application of chitin and chitosan in food packaging field is very limited. Yet, a large number of theoretical and computational studies have been reported that explain the structural and thermodynamic properties and other related characteristics that can form the basis of the application of chitin and chitosan-based nanocomposites in the food packaging area.

Nikolov et al. (2011) investigated the exoskeletons of arthropods and proposed a hierarchical model to study their elastic properties using computational calculations. The geometry of biomolecules obtained from atomistic calculations exhibited the existence of hydrogen bonding networks, providing high structural flexibility and change in conformation, to a greater extent in the case of the crystalline α-chitin. Thus, α-chitin was characterized by Born-Öppenheimer (BO) surface. However, the experiments could not easily determine the position of either hydrogen atoms or hydrogen bonds present in the network. Therefore, a hierarchical approach based on combination of empirical force field molecular dynamics (EFF-MD) calculations with self-consistent density functional tight binding (SC-DFTB) and DFT calculations was used in order to study the BO surface, as well as the ground-state geometry of the crystalline α-chitin. These calculations showed that lobster cuticles with some other biological nanocomposites had excellent mechanical properties that can be further implemented and modified for food packaging purposes.

Todde et al. have studied the adsorption of munition compounds (Todde et al. 2018) on cellulose chitin and cellulose triacetate using DFT calculations. Compounds such as 2,4-dinitroanisole (DNAN), 3-nitro-1,2,4-triazol-5-one (NTO), nitroguanidine (NQ), and 1,1-diamino-2,2-dinitroethene (FOX7) are insensitive munition compounds that can reduce the risk of explosions resulting from shock and exposure to high temperature. Sensitive munition compounds such as 2,4,6-trinitromethylbenzene (TNT) and 1,3,5-hexahydro-1,3,5-trinitro-1,3,5-triazine (RDX) have been recently replaced by insensitive munition compounds. The solubility of NTO and NQ is greater than that of TNT or RDX, which makes them spread and dissolve easily during contact with water. DNAN and TNT also exhibited comparable solubility. The natural polymers can adsorb these munition compounds to a large extent and hence, study of the interactions among the munition compounds and chitin, as well as cellulose, is carried out. DFT studies on the adsorption of munition compounds by cellulose Iα and cellulose Iβ, chitin, and cellulose triacetate have shown that all chitin and cellulose Iα adsorb the contaminants more readily than cellulose triacetate and cellulose Iβ. The results indicate that the adsorption of contaminants can be further investigated with chitin and other contaminants. This could also support the application of chitin nanocomposites in food packaging and other biomedical applications.

Molecular dynamics has also been used by Franca et al. to determine the structure and solubility of chitosan nanoparticles at different levels of deacetylation, and also study, as well as the effect of spatial distribution of N-acetyl groups in systems (Franca et al. 2011). Four different percentages of acetylation were considered: 0%, 40%, 60%, and 100%. It was confirmed that the systems in which the N-acetyl groups were uniformly distributed showed high flexibility and favored a relaxed two fold helix and five-fold helical structure. The systems in which the acetyl groups were restricted to a given region of the particles favored a two fold helix close to crystalline chitin. It was also concluded that systems with up to 40% acetylation showed moderate solubility. If the acetyl groups were nonuniformly distributed, stable aggregation was prominent. It was proved that systems with 60% and higher acetylation experienced similar aggregation when the acetylation was disturbed. The electrostatic force resulting from acetylation also affected the nature of solvation. The mobility and orientation of water around the polymer chains affected the stability of intramolecular hydrogen bonds which, in turn, affected the particle aggregation. In short, these studies showed the dependence of conformation and solubility with respect to pH and percentage of acetylation. As the percentage of acetylation was increased, flexibility and conformational interchangeability tended to be reduced. All simulations were done using the GROMOS 53A6 parameter, with the carbohydrate force field using the GROMACS 4.0.4 program. These factors that contribute to the flexibility of materials can show a large scope for the development of packaging materials.

The extraction of chitin from its natural sources mainly includes alkali and acidic treatments. Yu et al. demonstrated in 2014 that acidity affects the mechanical properties of the chitin protein interface due to the protonation and deprotonation of the ionizable groups present in chitin (Yudin et al. 2014). Using molecular dynamics, strength of adhesion of the protein on the chitin chain was examined. It was found that the level of protonation also affected the nature of adhesion. The mechanism of influence of acidity on the interface of chitin and protein was also addressed, which is useful for designing of specific biological materials. Steered molecular dynamics (SMD) simulations were executed using the massive parallel simulator LAMMPS, with CHARMM36 force fields.

Yudin et al. synthesized chitin fibrils reinforced chitosan biocomposites using coagulation methods (Yudin et al. 2014). The orientation of these chitosan molecules was studied using AMBER96 force field molecular simulations. The results confirmed that the orientation of chitosan molecules was dependent on the nanocrystalline structure of chitin. The nonparallel arrangement of chitosan to the chitin molecules confirmed weak interaction between them. In parallel or antiparallel orientation of chitosan, strong interaction was found to exist between chitin fibrils and the chitosan matrix. These findings were further supported by increased strength as well as Young's modulus.

Skovstrup et al. reported the flexibility studies of 1,4 linkage in the di-, tri- and tetra- saccharides of chitin and chitosan systems (Skovstrup et al. 2010). They found that the rotation of the glycosidic bond was dependent on the flexibility of neighboring linkage, and reported that the glucosamine N-acetylglucosamine linkage is more flexible than N-acetylglucosamine-N-acetylglucosamine linkage.

Faria et al. (2016) applied the molecular dynamics of chitin and chitosan nanoparticles to evaluate their various conformations and to focus on their solubility. This study was related to the influence of different arrangements of chitin and chitosan chains on the behavior of polymers in the aqueous medium. These arrangements affect interactions with solvent molecules via hydrogen bonding, swelling processes, etc. It was found that α- and β-chitin nanoparticles showed different degrees of swelling in water with α-chitin nanoparticles being more stable in water. Like chitin, α- and β-chitosan nanoparticles demonstrated similar behavior in solution. It was shown that the parallel orientation of chains in β-chitosan nanoparticles supported electrostatic interactions. The α-chitin with linear arrangement was found to exhibit greater stability and this has also been reflected in the swelling process. In the case of β-chitin, the total structure is lost with the linear arrangement. Molecular dynamics simulations were executed with GROMACS 4.5.3 program using GROMOS53a6 force field and were analyzed using VMD and Grace programs.

López-Chávez et al. carried out studies related to ion conductivity in chitosan membranes using molecular dynamics in 2005 (López-Chávez et al. 2005). The system considered for this study contained two polymeric chains, each with twelve amino groups, protonated chitosan monomers, one hydronium ion, one hydroxide ion, two hundred water molecules, and twelve sulfate ions. This structure was used to study the conductivity of hydronium and hydroxide ions. The molecular dynamics was carried out using COMPASS (condensed-phase optimized molecular potentials for atomistic simulation studies) force field. It was concluded that the mobility of charge carriers is dependent on the sulphate group attached to the amino group of the membrane. The amount of water and sulfates improved the ionic conductivity of the membrane. The results showed that these studies could be more reliable and economical for experimental design of new materials.

Van der Schueren et al. developed pH-sensitive polycaprolactone (PCL) nanofibers (Van der Schueren et al. 2013) and a blend of PCL/chitosan (20 wt%)/nitrazine composites. Computational calculations were carried out to investigate the electrostatic interactions between dyes and chitosan. The results showed that the interaction between chitosan and nitrazine was stronger, as compared to PCL. The molecular modeling studies were done using Gaussian09 software, with the M06-2X electronic structure method having a 6-311G (d,p) basis set. Figure 7.4 represents the optimized structures of the NY–PCL complexes (Figures 7.4A and 7.4C) and NY–chitosan complexes (Figures 7.4B and 7.4D) obtained using M06-2X/6-311G(d,p).

Prathab and Aminabhavi have conducted a wide range of molecular modeling for chitosan polymers in order to study the related surface, thermal, mechanical, and gas diffusion properties (Prathab

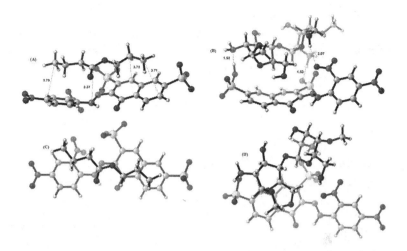

FIGURE 7.4 ((A) and (C)): NY–PCL complex optimized structure using M06-2X/6-311G(d,p), ((B) and (D)): NY–chitosan complex optimized structure using M06-2X/6-311G(d,p).

and Aminabhavi 2007). Molecular mechanics and molecular dynamics studies in two-dimensional periodic and three dimensional periodic condensed phases have been were carried out for the biopolymer. The surface energies and equilibrium structures obtained showed that the interior regions have a similar mass density in the bulk state. The specific volume of the polymers as a function of temperature has also been obtained. Polymer simulations have been done using MS modelling 3.1 and molecular mechanics and molecular dynamics was done with the DISCOVER package using COMPASS force field. The increased solubility of chitosan in comparison to chitin and cellulose was observed. In addition, chitosan was proved to be more hydrophilic due to the amino groups and swells in aqueous media. The free primary amino group provides conformational flexibility, which eases the possibility of modifying chitosan for creating biomaterial with antimicrobial activity. The diffusion coefficients of oxygen, nitrogen, and carbon dioxide in the polymers were also estimated. The presence of functional groups and hydrogen bonding reduced the mobility of polymer chains, thus reducing gas diffusion in functionalized chitosan polymers. In addition to these studies, many other computational works have also been reported describing the thermal, electrical, and physical properties of chitin- and chitosan-based polymers. From the results obtained, it can be inferred that chitin- and chitosan-based polymers and their nanocomposites can be efficiently utilized in the development of biomaterials, especially to meet the requirements of food packaging, when the material can be disposable, biodegradable, cost-effective, and nontoxic to nature.

REFERENCES

Debbaudt, Adrialla L, María Luján Ferreira, and María Elena Gschaider. 2004. "Theoretical and experimental study of M2+ adsorption on biopolymers. III. Comparative kinetic pattern of Pb, Hg and Cd." *Carbohydrate Polymers* no. 56 (3):321–332.

Dhanasekaran, Solairaj, Palanivel Rameshthangam, Suryanarayanan Venkatesan, Sanjeev Kumar Singh, and Sri Ramkumar Vijayan. 2018. "In vitro and in silico studies of chitin and chitosan based nanocarriers for curcumin and insulin delivery." *Journal of Polymers and the Environment* no. 26 (10):4095–4113.

Faria, Roberto Ribeiro, Renan Faria Guerra, Lourival Rodrigues de Sousa Neto, Luiz Frederico Motta, and Eduardo de Faria Franca. 2016. "Computational study of polymorphic structures of α-and β-chitin and chitosan in aqueous solution." *Journal of Molecular Graphics and Modelling* no. 63:78–84.

Franca, Eduardo F, Luiz CG Freitas, and Roberto D Lins. 2011. "Chitosan molecular structure as a function of N-acetylation." *Biopolymers* no. 95 (7):448–460.

Franca, Eduardo F, Roberto D Lins, Luiz CG Freitas, and Tjerk P Straatsma. 2008. "Characterization of chitin and chitosan molecular structure in aqueous solution." *Journal of Chemical Theory and Computation* no. 4 (12):2141–2149.

Gomes, José RB, Miguel Jorge, and Paula Gomes. 2014. "Interaction of chitosan and chitin with Ni, Cu and Zn ions: A computational study." *The Journal of Chemical Thermodynamics* no. 73:121–129.

Ho, Yi-Cheng, Zi-Xian Liao, Nilendu Panda, Deh-Wei Tang, Shu-Huei Yu, Fwu-Long Mi, and Hsing-Wen Sung. 2011. "Self-organized nanoparticles prepared by guanidine-and disulfide-modified chitosan as a gene delivery carrier." *Journal of Materials Chemistry* no. 21 (42):16918–16927.

Jayakumar, Rangasamy, Deepthy Menon, Koyakutty Manzoor, Shantikumar Vasudevan Nair, and Hiroshi Tamura. 2010. "Biomedical applications of chitin and chitosan based nanomaterials—A short review." *Carbohydrate Polymers* no. 82 (2):227–232.

Jensen, Frank. 2016. *Introduction to computational chemistry.* London, England: John Wiley and Sons.

Khor, Eugene, and Lee Yong Lim. 2003. "Implantable applications of chitin and chitosan." *Biomaterials* no. 24 (13):2339–2349.

Kim, Joong-Kwon, Do Hyeong Kim, Se Hun Joo, Byeongwook Choi, Aming Cha, Kwang Min Kim, Tae-Hyuk Kwon, Sang Kyu Kwak, Seok Ju Kang, and Jungho Jin. 2017. "Hierarchical chitin fibers with aligned nanofibrillar architectures: A nonwoven-mat separator for lithium metal batteries." *ACS Nano* no. 11 (6):6114–6121.

Lewars, Errol G. 2016. *Computational chemistry: Introduction to the theory and applications of molecular and quantum mechanics.* New York: Springer.

López-Chávez, Ernesto, José Manuel Martínez-Magadán, Raúl Oviedo-Roa, Javier Guzmán, Joel Ramírez-Salgado, and Jesús Marín-Cruz. 2005. "Molecular modeling and simulation of ion-conductivity in chitosan membranes." *Polymer* no. 46 (18):7519–7527.

Lü, Renqing, Zuogang Cao, and Guoping Shen. 2008. "Comparative study on interaction between copper (II) and chitin/chitosan by density functional calculation." *Journal of Molecular Structure: THEOCHEM* no. 860 (1–3):80–85.

Nikolov, Svetoslav, Helge-Otto Fabritius, Michal Petrov, Martin Friák, Liverios Lymperakis, Christoph Sachs, Dierk Raabe, and Jörg Neugebauer. 2011. "Robustness and optimal use of design principles of arthropod exoskeletons studied by ab initio-based multiscale simulations." *Journal of the Mechanical Behavior of Biomedical Materials* no. 4 (2):129–145.

Popuri, Srinivasa R, Yarramuthi Vijaya, Veera M Boddu, and Krishnaiah Abburi. 2009. "Adsorptive removal of copper and nickel ions from water using chitosan coated PVC beads." *Bioresource Technology* no. 100 (1):194–199.

Prabaharan, Mani. 2008. "Chitosan derivatives as promising materials for controlled drug delivery." *Journal of Biomaterials Applications* no. 23 (1):5–36.

Prasad Dhakal, Rabindra, Katsutoshi Inoue, Kazuharu Yoshizuka, Keisuke Ohto, Mariko Yamada, and Sumito Seki. 2005. "Solvent extraction of some metal ions with lipophilic chitin and chitosan." *Solvent Extraction and Ion Exchange* no. 23 (4):529–543.

Prathab, Baskar, and Tejraj M Aminabhavi. 2007. "Molecular modeling study on surface, thermal, mechanical and gas diffusion properties of chitosan." *Journal of Polymer Science Part B: Polymer Physics* no. 45 (11):1260–1270.

Rinaudo, Marguerite. 2006. "Chitin and chitosan: properties and applications." *Progress in Polymer Science* no. 31 (7):603–632.

Skovstrup, Søren, Signe Grann Hansen, Troels Skrydstrup, and Birgit Schiøtt. 2010. "Conformational flexibility of chitosan: a molecular modeling study." *Biomacromolecules* no. 11 (11):3196–3207.

Svetlana, Jeremić, Thu Hien Tran, Zoran Marković, Chinh Thi Ngo, and Duy Quang Dao. 2018. "Insight into interaction properties between mercury and lead cations with chitosan and chitin: density functional theory studies." *Computational and Theoretical Chemistry* no. 1138:99–106.

Todde, Guido, Sanjiv K Jha, Gopinath Subramanian, and Manoj K Shukla. 2018. "Adsorption of TNT, DNAN, NTO, FOX7, and NQ onto cellulose, chitin, and cellulose triacetate. Insights from density functional theory calculations." *Surface Science* no. 668:54–60.

Van der Schueren, Lien, Thierry De Meyer, Iline Steyaert, Özgür Ceylan, Karen Hemelsoet, Veronique Van Speybroeck, and Karen De Clerck. 2013. "Polycaprolactone and polycaprolactone/ chitosan nanofibres functionalised with the pH-sensitive dye Nitrazine Yellow." *Carbohydrate Polymers* no. 91 (1):284–293.

Young, David. 2004. *Computational chemistry: a practical guide for applying techniques to real world problems.* New York: John Wiley & Sons.

Yu, Zechuan, Zhiping Xu, and Denvid Lau. 2014. "Effect of acidity on chitin–protein interface: a molecular dynamics study." *BioNano-Science* no. 4 (3):207–215.

Yudin, Vladimir E, Irina P Dobrovolskaya, Igor M Neelov, Elena N Dresvyanina, Pavel V Popryadukhin, Elena M Ivan'kova, Vladimir Yu Elokhovskii, Igor A Kasatkin, Boris M Okrugin, and Pierfrancesco Morganti. 2014. "Wet spinning of fibers made of chitosan and chitin nanofibrils." *Carbohydrate Polymers* no. 108:176–182.

CHAPTER 8

Novel Approaches for Chitin/Chitosan Composites in the Packaging Industry

Victor Gomes Lauriano Souza, João Ricardo Afonso Pires, and Carolina Rodrigues

MEtRICs, Departamento de Ciências e Tecnologia da Biomassa,
  Faculdade de Ciências e Tecnologia, FCT
Universidade Nova de Lisboa, Campus de Caparica
Almada, Portugal

Isabel Coelhoso

LAQV-REQUIMTE, Departamento de Química, Faculdade de
  Ciências e Tecnologia
Universidade Nova de Lisboa, Campus de Caparica
Almada, Portugal

Ana Luisa Fernando

MEtRICs, Departamento de Ciências e Tecnologia da Biomassa,
  Faculdade de Ciências e Tecnologia, FCT
Universidade Nova de Lisboa, Campus de Caparica
Almada, Portugal

8.1 INTRODUCTION

Chitin- and chitosan-based composites play an important role in food packaging applications; they are renewable, nontoxic and biodegradable feed stocks, capable of forming membranes, gels, or films, thus replacing traditional environmentally unfriendly petroleum-based materials. However, the following hindrances limit large scale use of these polysaccharides: cost, processing and performance properties (i.e., reduced barrier, thermal and mechanical properties) (Souza et al. 2018b). Various approaches to chitin/chitosan composites in the food packaging industry are emerging as a mean to surpass the drawbacks of these polymers. This chapter shows the current trends and the future challenges to chitin/chitosan composites in food packaging.

8.2 CHITIN AND CHITOSAN PRODUCTION: NOVEL EXTRACTION METHODS AND BIOLOGICAL PRODUCTION

Chitosan is commercially produced from chitin. It is naturally present in a variety of species, such as green algae and crustaceans (in their exoskeletons). Currently, shells from crustaceans (e.g., shrimp, crabs) are the main sources of chitin. Chitosan production involves several chemical processes, including decalcification, deproteinization, decolourization, and deacetylation. A novel approach in the extraction of chitosan is the use of Deep Eutectic Solvents (DES) and a new class of green solvents (Zdanowicz et al. 2018). These biodegradable solvents are obtained simply and their properties can be shaped according to their applications (Zdanowicz et al. 2018). These solvents were successfully used to dissolve chitin while avoiding alkali treatment, traditionally used to produce high purity chitosan by deacetylation (Figure 8.1). The green solvents are found to be effective for removing proteins and minerals with a yield higher than the chemically prepared chitosan. Other environmentally safe alternatives for preparing chitin and chitosan include biological methods, in which lactic acid

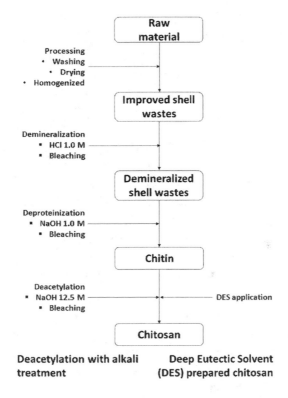

Processing
• Washing
• Drying
• Homogenized

Demineralization
■ HCl 1.0 M
■ Bleaching

Deproteinization
■ NaOH 1.0 M
■ Bleaching

Deacetylation
■ NaOH 12.5 M
■ Bleaching

DES application

Raw material

Improved shell wastes

Demineralized shell wastes

Chitin

Chitosan

Deacetylation with alkali treatment   Deep Eutectic Solvent (DES) prepared chitosan

FIGURE 8.1   Dissolution of chitin with Deep Eutectic Solvents (DES) in substitution of the deacetylation via alkali treatment.

(produced though microorganism fermentation) is enrolled in the demineralization step and proteases perform the deproteinization, as well as efficient and speedy methods (e.g., microwave, enzymatic, or photochemical methods) (El-Knidri et al. 2018).

Chitin obtained from shellfish waste may be limited in scale due to volatile resource availability, the possible residual presence of allergens and/or contaminants which requires extra purification/refining steps, and increased production cost and low degree of deacetylation (Ozkan and Rorrer 2017). Biological production of chitin and chitosan is being considered as an alternative process to ensure availability and to reduce the problematic allergen issue. Algae are also considered to be significant sources for production of chitin and chitosan, along with other added value compounds, such as lipids. The centric diatoms (e.g. *Cyclotella* sp. or *Thalassiosira* sp.) produce β-chitin, which can be extruded from cells (Chiriboga and Rorrer 2019). Zygomycetes can also produce chitosan through a process involving synthesis of chitin by chitin synthase, which is subsequently converted to chitosan by chitin deacetylase. The fungal biomass can be produced via fermentation at low cost, and since no demineralization step is required, chitosan extraction can be classified as a green process. Additionally, fungal chitosan has a medium-low molecular weight, compared with the chitosan extracted from crustaceans, along with higher bioactivity (Abdel-Gawad et al. 2017). The characteristics and amount of chitosan removed from fungi (including mushrooms and mushroom wastes), depend on several factors, namely, type of fermentation and microorganism, type of process, and operational parameters such as pH, temperature, extraction time or composition of culture (Bilbao-Sainz et al. 2017). The use of synthetic media, which is costly, can be substituted by agro-industrial wastes, promoting the bioeconomy and circular economy and reducing environmental impacts (Vendruscolo and Ninow 2014). Fungi are used in different industrial processes, e.g., baking industry, production of antibiotics, resulting in tons of fungal waste biomass annually. Thus, chitosan extraction from these industrial wastes is another option. Current research is also focusing on the increment of the chitosan content in the cell walls of fungi through strain improvement and metabolic engineering (Ghormade et al. 2017).

8.3 IMPROVEMENT OF CHITOSAN PROPERTIES

The incorporation of nanoscale reinforcements (e.g., montmorillonite, nanocellulose, metal oxide nanoparticles) in chitosan films is a strategy to improve the poor mechanical and barrier properties of this bio-based polymer. The nanomaterials can interact with the polymer, changing its structure and properties. Montmorillonite (MMT) mineral clay is widely available at low cost, and when applied to strengthen chitosan-based composites, the results showed improved mechanical properties and increased UV and oxygen barrier in the films (Souza et al. 2018a).

Cellulose nanofibers and cellulose nanocrystals are appealing reinforcements because of high compatibility with chitosan. The electrostatic association and hydrogen bonds between nanocellulose with large length–diameter ratios and chitosan molecules causes the formation of an interactive network structure that provides an increment in the film's crystallinity (Mao et al. 2019). Nanofillers, such as nanoscale metal oxides (e.g., ZnO, SiO_2), may, in addition to reinforcement capacity, confer antimicrobial properties or UV blocking characteristics (Tian et al. 2019).

Another approach for improving the characteristics of chitosan films is the incorporation of lipids (e.g., waxes or resins), which can impart hydrophobicity to the film and reduce its moisture content (Galus and Kadzińska 2015). Research on chitosan also focuses on the identification of the active biocompounds that confer better antioxidant and antimicrobial capacities to edible films (Souza et al. 2018c). The advantage of these active packages is the avoidance of taste transfer to the food, which reduces organoleptic changes. Additional work must still be done to understand the interactions between chitosan and bioactive compounds, so that the film's bioactivity and physical properties can be optimized (Souza et al. 2017). Research is also needed to overcome the challenge of keeping the performance of essential oil/extracts in films during the production process of polymers, which generally demand high temperatures (Atarés and Chiralt 2016). Nevertheless, it is also important to have information on the biodegradability of these bionanocomposites, as well on as their toxicity and ecotoxicity (Souza et al. 2018b).

The intrinsic reactive groups of chitosan, namely –OH and $-NH_2$, allow the chemical modification of chitosan, enhancing its application potential. The amino groups present in chitosan react with carbonyl compounds via imine functionalization resulting in chitosan-based Schiff bases which are of importance for certain food packaging applications (Figure 8.2).

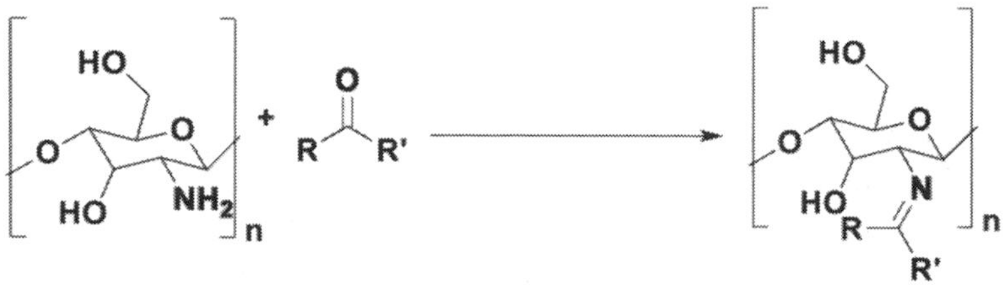

FIGURE 8.2 Production of chitosan-based Schiff bases from chitosan and aldehydes/ketones.

Chitosan-based Schiff bases have shown antimicrobial activities as powders/whiskers/films/membranes and interestingly, present antimicrobial properties superior to those of pristine chitosan (Antony et al. 2019). Moreover, the antimicrobial action of chitosan-based Schiff bases can be augmented through incorporation of metal ion nanoparticles. Some chitosan-based Schiff bases have also shown antioxidant activity, thereby improving the functional properties of bare chitosan.

Improvements in the functional properties of chitosan was also reported when this biopolymer was combined with other macromolecules such as starches (Luo and Wang 2014), microbial polysaccharides (Freitas et al. 2014; Ferreira et al. 2016), or proteins (Kurek et al. 2014). The enhancement of the properties was attributed to electrostatic interactions. The amino groups present in chitosan (protonated) interact with the other biopolymers with negatively charged groups (Luo and Wang 2014). In blends, insoluble complexes can also be formed between polymers. Thus, in order to overcome such constraint, bilayer systems are an alternative to blend films, and some reports have shown that such materials have reduced permeability to water vapor (Kurek et al. 2014).

8.4 APPLICATION AS COATING

Chitosan can also be used as a coating material, a layer that can be applied directly to the surface of food, as an edible coating, or on the surface of packaging materials to functionalize them (Vasile 2018). With the advance of nanotechnology, a novel concept has appeared, called nanocoatings, which consist of an ultrathin nanolayer with a thickness (below 100 nm) placed on surfaces. This type of coatings have the advantage of not modifying the materials' surface topography while can add physical and chemical functions to the surface, such as change gas-barrier properties, surface hydrophobicity, or conductive properties (Vasile 2018). The deposition of liquid dispersion (applied, for example, with a paintbrush or via spraying) or of molten compounds which form film directly on the food surface generate the edible coating. Ripening, as well as water loss problems in fruits and vegetables, can be retarded with the use of coatings, reducing decay in these products (Aranaz et al. 2009). Coatings can also improve the quality of meat products by retarding moisture loss, enhancing product appearance, diminishing discoloration and oxidation, and serving as a carrier of food additives (Pires et al. 2018).

8.5 SCALE-UP PRODUCTION

Chitin and chitosan are produced worldwide. In Iceland, Primex produces the "ChitoClear®", to be used by the food packaging industry. In Norway, Norwegian Chitosan commercializes chitin (Norlife) and chitosan (Kitoflok™), for various applications including food and beverages. G.T.C. Bio Corporation (China) trades different grades of chitin and chitosan. Yet, most of the data available on chitosan films and coatings applications are related to their production via casting methods, mainly used in laboratory scale. Plasticization by thermomechanical treatments is a potential alternative to the customary casting method, allowing production of chitosan films with good mechanical properties (Galvis-Sánchez et al. 2018). Green solvents, such as DES and natural deep eutectic solvents (NADES), are suitable for use as plasticizers on the production of thermoplastic chitosan films.

Various approaches to chitin/chitosan composites in the food packaging industry are emerging as a means of solving some of the drawbacks of chitosan stand-alone polymers. In this paper, the current trends and future challenges of chitin/chitosan composites in the food packaging industry were presented. The following specific topics were outlined: new and more sustainable processes for extracting chitin, biological production of chitin and chitosan, improvement of chitosan's properties through chemical modification, incorporation of nanofillers and bioactive agents or as blends and bilayers with other biopolymers, applications as coatings, and scaling-up the process.

REFERENCES

Abdel-Gawad, K. M., Awatief, F. H., Mustafa, A. F., and Gomaa, M. 2017. Technology Optimization of Chitosan Production from *Aspergillus niger* Biomass and Its Functional Activities. *Food Hydrocolloids* 63: 593–601. doi:10.1016/j.foodhyd.2016.10.001.

Antony, R., Arun, T., and Manickam, S. T. D. 2019. A Review on Applications of Chitosan-Based Schiff Bases. *International Journal of Biological Macromolecules* 129: 615–33. doi:10.1016/j.ijbiomac.2019.02.047.

Aranaz, I., Mengíbar, M., and Harris, R. 2009. Functional Characterization of Chitin and Chitosan. *Current Chemical Biology* 3 (2): 203–30. www.ingentaconnect.com/content/ben/ccb/2009/00000003/00000002/art00009.

Atarés, L., and Chiralt, A. 2016. Essential Oils as Additives in Biodegradable Films and Coatings for Active Food Packaging. *Trends in Food Science and Technology* 48: 51–62. doi:10.1016/j.tifs.2015.12.001.

Bilbao-Sainz, C., Chiou, B. S., Williams, T., Wood, D., Du, W. X., Sedej, I., Ban, Z., Rodov, V., Poverenov, E., Vinokotur, Y., and McHugh, T. 2017. Vitamin D-Fortified Chitosan Films from Mushroom Waste. *Carbohydrate Polymers* 167: 97–104. doi:10.1016/j.carbpol.2017.03.010.

Chiriboga, O., and Rorrer, G. L. 2019. Phosphate Addition Strategies for Enhancing the Co-production of Lipid and Chitin Nanofibers during Fed-Batch Cultivation of the Diatom *Cyclotella sp. Algal Research* 38: 101403. doi:10.1016/j.algal.2018.101403.

El-Knidri, H., Belaabed, R., Addaou, A., Laajeb, A., and Lahsini, A. 2018. Extraction, Chemical Modification and Characterization of Chitin and Chitosan. *International Journal of Biological Macromolecules* 120: 1181–89. doi:10.1016/j.ijbiomac.2018.08.139.

Ferreira, A. R. V., Torres, C. A. V., Freitas, F., Sevrin, C., Grandfils, C., Reis, M. A. M., Alves, V. D., and Coelhoso, I. M. 2016. Development and Characterization of Bilayer Films of FucoPol and Chitosan. *Carbohydrate Polymers* 147: 8–15. doi:10.1016/j.carbpol.2016.03.089.

Freitas, F., Alves, V. D., Reis, M. A., Crespo, J. G., and Coelhoso, I. M. 2014. Microbial Polysaccharide-Based Membranes: Current and Future Applications. *Journal of Applied Polymer Science* 131 (6): 40047. doi:10.1002/app.40047.

Galus, S., and Kadzińska, J. 2015. Food Applications of Emulsion-Based Edible Films and Coatings. *Trends in Food Science and Technology* 45 (2): 273–83. doi:10.1016/j.tifs.2015.07.011.

Galvis-Sánchez, A. C., Castro, M. C. R., Biernacki, K., Gonçalves, M. P., and Souza, H. K. S. 2018. Natural Deep Eutectic Solvents as Green Plasticizers for Chitosan Thermoplastic Production with Controlled/Desired Mechanical and Barrier Properties. *Food Hydrocolloids* 82: 478–89. doi:10.1016/j.foodhyd.2018.04.026.

Ghormade, V., Pathan, E. K., and Deshpande, M. V. 2017. Can Fungi Compete with Marine Sources for Chitosan Production? *International Journal of Biological Macromolecules* 104: 1415–21. doi:10.1016/ j.ijbiomac.2017.01.112.

Kurek, M., Galus, S., and Debeaufort, F. 2014. Surface, Mechanical and Barrier Properties of Bio-based Composite Films Based on Chitosan and Whey Protein. *Food Packaging and Shelf Life* 1 (1): 56–67. doi:10.1016/j.fpsl.2014.01.001.

Luo, Y., and Wang, Q. 2014. Recent Development of Chitosan-Based Polyelectrolyte Complexes with Natural Polysaccharides for Drug Delivery. *International Journal of Biological Macromolecules* 64: 353–67. doi:10.1016/j.ijbiomac.2013.12.017.

Mao, H., Wei, C., Gong, Y., Wang, S., and Ding, W. 2019. Mechanical and Water-Resistant Properties of Eco-Friendly Chitosan Membrane Reinforced with Cellulose Nanocrystals. *Polymers* 11 (1): 166. doi:10.3390/polym11010166.

Ozkan, A., and Rorrer, G. L. 2017. Effects of CO_2 Delivery on Fatty Acid and Chitin Nanofiber Production during Photobioreactor Cultivation of the Marine Diatom *Cyclotella sp. Algal Research* 26: 422–30. doi:10.1016/j.algal.2017.07.003.

Pires, J. R. A., Souza, V. G. L., and Fernando, A. L. 2018. Chitosan/ Montmorillonite Bionanocomposites Incorporated with Rosemary and Ginger Essential Oil as Packaging for Fresh Poultry Meat. *Food Packaging and Shelf Life* 17: 142–49. doi:10.1016/j.fpsl. 2018.06.011.

Souza, V. G. L., Fernando, A. L., Pires, J. R. A., Rodrigues, P. F., Lopes, A. A. S., and Braz-Fernandes, F. M. 2017. Physical Properties of Chitosan Films Incorporated with Natural Antioxidants. *Industrial Crops and Products* 107: 565–72. doi:10.1016/j.indcrop. 2017.04.056.

Souza, V. G. L., Pires, J. R. A., Rodrigues, P. F., Lopes, A. A. S., Fernandes, F. M. B., Duarte, M. P., Coelhoso, I. M., and Fernando, A. L. 2018a. Bionanocomposites of Chitosan/Montmorillonite Incorporated with Rosmarinus Officinalis Essential Oil: Development and Physical Characterization. *Food Packaging and Shelf Life* 16: 148–56. doi:10.1016/j.fpsl.2018.03.009.

Souza, V. G. L., Ribeiro-Santos, R., Rodrigues, P. F., Otoni, G. C., Duarte, M. P., Coelhoso, I. M., and Fernando, A. L. 2018b. Nanomaterial Migration from Composites into Food Matrices. In *Composite Materials for Food Packaging*, eds. G. Cirillo, M. A. Kozlowski, and U. G. Spizzirri, 395–430. Beverly: Scrivener Publishing LLC.

Souza, V. G. L., Rodrigues, P. F., Duarte, M. P., and Fernando, A. L. 2018c. Antioxidant Migration Studies in Chitosan Films Incorporated with Plant Extracts. *Journal of Renewable Materials* 6 (5): 548–58. doi:10.7569/JRM.2018.634104.

Tian, F., Chen, W., Wu, C. E. E., Kou, X., Fan, G., Li, T., and Wu, Z. 2019. Preservation of Ginkgo Biloba Seeds by Coating with Chitosan/Nano-TiO$_2$ and Chitosan/Nano-SiO$_2$ Films. *International Journal of Biological Macromolecules* 126: 917–25. doi:10.1016/j. ijbiomac.2018.12.177.

Vasile, C. 2018. Polymeric Nanocomposites and Nanocoatings for Food Packaging: A Review. *Materials* 11 (10): 1834. doi:10.3390/ ma11101834.

Vendruscolo, F., and Ninow, J. L. 2014. Apple Pomace as a Substrate for Fungal Chitosan Production in an Airlift Bioreactor. *Biocatalysis and Agricultural Biotechnology* 3 (4): 338–42. doi:10.1016/j. bcab.2014.05.001.

Zdanowicz, M., Wilpiszewska, K., and Spychaj, T. 2018. Deep Eutectic Solvents for Polysaccharides Processing. A Review. *Carbohydrate Polymers* 200: 361–80. doi:10.1016/j.carbpol.2018.07.078.

Index